Careers in Space

Also by Otto O. Binder
Victory in Space

CAREERS IN SPACE

Otto O. Binder

WILDSIDE PRESS

Careers in Space

Published by Wildside Press LLC
wildsidepress.com

With confidence, this book is dedicated to those young Americans of today destined to make the space history of tomorrow

FOREWORD

This book is a tour de force, a thorough examination of the aerospace industry and the national space program which should have encyclopedic value for anyone interested in the broad sweep of space activities across the nation.

College students completing, or having recently completed, a degree in science or engineering will find this book's profiles of space industries invaluable in helping them identify those firms needing people in their degree fields.

High-school students will find in this book a wealth of information about everything from available scholarships to descriptions of such pioneering fields of study as exobiology and bionics. And the comprehensive detail which Mr. Binder presents will help students to become more knowledgeable about the nation's rapidly growing complex of industries related to manned and unmanned space exploration.

Some of the issues which Mr. Binder discusses are frankly controversial, and not everyone will agree—or indeed, is expected to agree—with all his opinions. But Careers in Space is not primarily a thesis book. Its strength lies rather in the depth and breadth of the data it contains. It is impressive, quite simply, because it is so enormously informative and useful.

Paul L. Gardner
Educational Services Specialist
Office of Educational Programs and Services
NATIONAL AERONAUTICS AND SPACE ADMINISTRATION

FOREWORD

Since October 4, 1957, when the Soviet Union launched its first Sputnik, we have been living in the Space Age. It is a truly exciting time in which to be alive. Compared to other periods of history, our knowledge of the universe is increasing at an extremely rapid rate, and the end is nowhere in sight. Each new fact that is learned and each discovery that is made leads us to still more information. After being confined for centuries to the planet Earth, man, in the first few years of the Space Age, has made remarkable progress in expanding his horizons.

In the Space Age no country can ever be satisfied with its past accomplishments, remarkable as they are. Too much will always remain to be done. The possibilities are as infinite as the space above the earth. The challenge of space is the greatest we have ever faced, both as a country and as members of the human race.

How well we meet the challenge depends on how wisely we use our resources, particularly that most important of all resources, manpower. No matter how large a budgetary appropriation is made for an important space project, its ultimate success or failure will depend on the skill and dedication of the people who do the work. The challenge of the Space Age demands that we locate our best scientific talent, train it and use it wisely.

In a free country such as ours young people can choose any career they wish within the limits of their talents and abilities. Many who are well equipped for a career in the new, but vitally important, field of space exploration, or astronautics, choose to do something else. Astronautics is a new field. Its requirements, job opportunities and rewards *are not nearly as well known as those of law or medicine, for example. If they were, more young people would be preparing for careers in astronautics.*

Otto Binder in Careers in Space *presents this information in detail to the high school and college student, along with practical advice on such things as choosing a school and a course of study, obtaining financial aid through scholarships or student loans, and where to look for the right position after graduation.*

The young man or woman who doesn't graduate from a four-year college is not excluded from a career in one of our many space endeavors. There are responsible, exciting jobs open to those who qualify as technicians. Junior colleges, vocational schools and the many training opportunities offered by the armed services can all lead to a career in space.

The important thing is that young people realize we have entered the Space Age. Whether we move boldly ahead or fall behind depends upon them. They must supply the know-how that takes us from an exploration of the moon to the exploration of more distant parts of the universe. The problems to be solved are many and difficult, but they can be solved. Those who choose a career in astronautics will have the satisfaction of helping their country prepare for the future. In whatever capacity they serve, civilian or military, engineer, scientist or technician, they will be using their talents to meet the greatest challenge we have yet faced.

GENE GURNY
Major, USAF
Director, USAF Book Program
Office of Information

PREFACE

What is America's greatest wealth? Oil underground, gold bullion in Fort Knox, stockpiles of uranium? All such tangible assets put together are petty cash compared to our most priceless treasure.

President Kennedy has defined it very clearly: "The *human mind* is our most fundamental resource."

Throughout history, the human mind has been the fountainhead of discovery and invention, turning ideaware into hardware. We have heard of the Space Race. Even more important to our nation's future status is the Brain Race. Every mind counts if our country is to remain first-rate.

But a strange paradox has arisen in America with the opening of the Space Age. In the face of a mindpower shortage that is particularly strangling the newborn astronautics industry, engineering enrollments in colleges have fallen off for five straight years. Incredibly, the time-honored law of supply and demand has broken down. Everyone seems aware that more scientists and engineers are desperately needed, yet fewer are graduating to take up those vital careers.

Why?

Nobody quite knows the baffling answer, and it has brought up a serious dilemma. "One of the most critical problems facing this nation," stated the President recently, "is the inadequacy of the supply of scientific and technical manpower [for] this country's research and development efforts in the near future."

Sterling M. McMurrin, who recently retired as Federal Commissioner of Education, bluntly stated during his farewell speech that America was wasting its talent "in a most disgraceful manner." He added that the greatest threat facing our future as a people was the "undemocratic conception of equality" regardless of ability, and the "irrational lack of respect for reason."

Dr. McMurrin's successor, Dr. Francis Keppel (Dean of Harvard's Graduate School of Education), is no less vehement in deploring the state of our educational system, and is seeking widespread reforms. He is the voice of many educators who warn that we will lose the Science Race unless every American youth with technological talent is found, trained and utilized. At the minimum, they add, we must double our rate of graduating engineers and scientists for the rest of this decade. There must be a full-scale mental mobilization from the only source available—our schools.

This means *you*, if you are a student in high school or college today with an aptitude for science. You are the intellectual recruits whose brainpower will guarantee that our nation remains a world power and the champion of peace and freedom. Your mental thrust is the only propulsion system that can really conquer space frontiers.

PLAIN TALK

Perhaps you should be forewarned that this book is to be hard-hitting. It is not a time for pulling punches or glossing over the issues. You junior citizens of today hold the future of the United States in your hands. And it must be plainly said that if you pre-college students and university undergrads fail to make a massive swing toward professional careers in science technology, then you will undoubtedly condemn yourselves to live your later lives in a second-string nation, trampled down by the rest of the stampeding world.

Notice the explicit phrase *science technology* above, meaning science or engineering *specifically* and nothing else. To be brutally candid, becoming football stars won't help. Matriculating as doctors, lawyers or journalists won't help. Joining the ranks of salesmen, business executives or architects won't help. None of those occupations, respectable and time-honored though they are, can help us keep up during the *technological revolution* into which the world has been plunged.

One major prejudice must be overcome. In a study made by the Horace Mann-Lincoln Institute of School Experimentation, under the title "Talented Youth Project," it was found by a poll among students that the general "climate" in American schools still extolls the "cheerful mediocrity" of football heroes and penalizes the serious student trying to improve his brain. Mental brilliance is equated with a "freakish quality" in the superior student of science.

What is really freakish is this *attitude*. In the rest of the world—Europe, Russia, Asia, South America, even Africa—the learned mind is honored and respected for its true worth.

Perhaps this anti-intellectualism was handed down to you junior citizens by your elders. Perhaps it is mainly your own slothful fault. Wherever the blame lies, do some hard thinking about it. What kind of attitude are you going to hand to the generation *after* yours? And what kind of *America?* It will be an America of "freakish quality"—to the rest of the world—because its citizens have a backward attitude toward training their brains.

To be plain, you are academic "squares" in the eyes of the other pro-intellectual nations around this earth. It is America that is out of step, not the world. If you have a mind of your own, use it. Being mentally anemic or flabby between the ears is no different from letting your body become anemically skinny or a flabby-muscled tub of lard. Worse yet, wasting a good brain is precisely the same as throwing away a chest of jewels. The cheerful-minded athlete will be a drug on the future market as America's demand for the champion-minded creator grows.

And remember that only becoming scientists and engineers will help in building up the technological *brainthrust* your country badly needs.

This is not to extoll science technology as the epitome of a thinking nation's goals. The humanities, music, literature—all the liberal arts—are undeniably cultural cream that is also important to our way of life. But things are twisted if it seems this book is promoting an "overbalance" of technically oriented minds. The simple truth is that America has *fallen behind* in recruiting laboratory talent. The onrush of events has left us clinging to horse-and-buggy standards in education while thundering rockets hurl vehicles into space. If we fear being overrun by too many techni-eggheads, let us look at the record. Out of a total roster of 515,000 college and university graduates in 1962, diplomas were granted to 56,000 in the engineering professions and physical sciences— namely about *one-tenth*. Doubling that amount, as recommended by educators, will hardly disturb America's traditional academic balance between the "materialistic" and the "esthetic."

However, let us put the horse before the cart. None of the esthetic arts can flower in the first place unless we have freedom from aggression. But aggression will be invited if our slippage in progress grows worse. As General Schriever has said: "Our first business —like it or not—is *survival*."

SOVIET CHALLENGE

It is hardly a secret that Communist Russia has declared scientific war. Hostilities formally opened in 1957 when Soviet space guns fired the Sputnik into human history. Along with it came a great cold-water shock—an unmistakable challenge to America's technological mastery, where we had thought we were supreme.

With the Sputnik the ICBM was spawned—the rocket-powered atomic bomb able to span the earth. But many experts do not fear the possibility of hot *nuclear* war as much as a cold-war *Battle of Brains*. In such a ten- or twenty-year bloodless struggle for world leadership, the battle-lines will be drawn in our campuses and classrooms, in our laboratories and engineering plants. The hard-won victory will come to the nation with the biggest stockpile of breakthrough bombs, the nation that creates the greatest scientific fallout.

We might, somehow, survive a nuclear war's massive mayhem. But losing the laboratory war will be even more devastating and will mean total defeat in the forthcoming age of applied intelligence.

Much of our country's long-term struggle to win the science showdown with Russia will focus on our space program. Yet there is no such thing as "space science" in itself. It is compounded of a dozen other physical sciences—astronomy, physics, chemistry, biology and down the line. Almost any field of R&D (Research and Development) the student of today chooses will inevitably give him a part, large or small, in the astronautics endeavor.

The National Aeronautics and Space Administration (NASA) needs many more hard-to-find space personnel each year for its mushrooming projects. Aerospace firms have had help-wanted ads out day and night and Sundays, begging for technical people. This shortage of engineering and scientific manpower is pinching America where it hurts the most. You in the classrooms today must cure the scientific anemia sapping America's strength.

The Science Race is real, it is earnest—at least, it is to the Soviets. Their stated goal is to "bury" us, if not by war then in a scientific landslide that will slowly smother us through the years ahead. Unfortunately, past security and confidence keep us from fully realizing the future danger. Dr. Martin A. Elliot, Vice President of the Illinois Institute of Technology, blames "America's complacent attitude toward progress in education" for the nation's "toe-to-toe struggle with Russia for technical and scientific supremacy."

From the business world too, besides academic and government quarters, the same alarm rings. M. G. O'Neil, President of the General Tire and Rubber Company (deep in aerospace projects with its Aerojet-General Division), not only says that "backing away from the Space Age would be denying us the great gift of progress," but adds, "If Russia beats us to the punch, we shall have only ourselves to blame when the Hammer and Sickle tears the Stars and Stripes to shreds."

That's about as straightforward as you can get it from one of America's leading industrialists. The message is stark and simple—only you in today's schools can form the USA's brain battalion of tomorrow and stop that Communist coup to which the Kremlin seems implacably pledged.

True, no one can positively state beyond a shadow of a doubt whether such a steadfast Soviet plan exists, or can succeed if it does exist. But how can America take a chance? By guarding against the worst, we lose nothing either way. By assuming the best—and getting caught—we would lose everything. The choice does not seem difficult to make under these circumstances of it's-them-or-us.

Don't let yourself *and* your country down. Science and space want *YOU*.

THE SCIENCE FRONTIER

But there are other and better motives than just *reacting* to something done by another nation. Being *forced* to take steps is a negative way of making progress. Even if the Russians didn't exist, we should still *want* to tackle the secrets of the universe. This is the true inspiration that should guide you of the upcoming generation —if you have the proper ability—to seek your lifework in scientific search. You have a variety of worthwhile motives for choosing to march into the frontiers of uncharted knowledge.

Do you like *challenge?*

Then you can be part of the greatest mettle-testing task in mankind's history in the coming conquest and exploration of space. Though as an earthbound scientist, engineer or technician you may not enjoy the thrill of first placing human footprints on other worlds, you will nevertheless bask in reflected glory as one of the "men behind the spacemen."

Do you dream of *fame?*

Worlds of science as well as space remain to be explored for lifetimes ahead. The pace of technology will speed up until breakthroughs come almost daily and the discovery of the century rolls around every decade. "How I envy those beginning their careers," sighed an elderly scientist. "They will inherit a wonderful age beyond human imagination in which young researchers will win honors at their peak, in their twenties and early thirties. Space science is a *young* man's game."

Does the *pioneering* spirit in you demand satisfaction?

"Landing a manned expedition on the moon will be a beginning," stated President Kennedy, "not an end . . . Flights to Mars and Venus are being contemplated [and] spacecraft to investigate the fiery regions near Mercury and the sun . . . and the remote reaches of the solar system . . . " Many of you about to start astronautics careers will, within your lifetimes, become members of expeditions, camps and colonies on exotic sister worlds of earth. As with the Mercury astronauts, every space crew will out of necessity be composed of technically trained personnel.

But leaving the lofty for the practical, do you seek *job security?*

The present-day technical manpower shortage will be eased, but never fully overcome, simply because it will be humanly impossible to catch up with the science technology explosion. If ever any young men were "guaranteed" a college-to-grave career, it is your generation today. Space exploration alone—aside from all non-astronautics beehives of R&D—will keep you from ever knowing what unemployment is, even if medical miracles extend your life for centuries. Space is here to stay.

On the score of your *earnings*, there already are signs of a "salary explosion" as NASA, the government, industry and academic institutions all frantically bid for the services of the "uncommon man" with a powerhouse between his ears. America will also—in the sense of competing in the Space Race—have to "meet" the Russian pay-scale, which is highest of all for scientists and engineers, five to ten times that of their national average for other trades.

Boredom, a penalty many Americans must pay in humdrum jobs, can never strike you during a lifelong series of Missions Into Mystery. *Honors* will be strewn in your path, as time, which changes all things, elevates space heroes above football heroes and science stars above singing stars.

Hard work will not be denied you in your space job, for after doing the impossible, you will be given the inconceivable. *Responsibility* will be yours too, particularly if you work on some phase of a manned project, where any faulty device will mean life and death to spacemen. All these will make demands on your *courage*, your *toughness of mind*, your *confidence*.

And these are still the *good* things about your science career, not drawbacks. Perhaps the highest degree of true happiness comes to the man who proves himself tougher than the toughest challenge thrown before him. Beyond the mere "fringe benefits" of good earnings, job security and honors comes the rare coin of *self-satisfaction*.

No other field of human endeavor can bring man the unsurpassable triumphs of an Einstein or an Edison, who discovers or designs that which no other human being before him has ever conceived. That is your true payoff.

YOUR SPACE GUIDE

To be specific, this book will first look over the educational field and astronautics industry in a broad sweep, then go into detail in order to offer definite guidelines for you, the space-science student. Statistics, tables and most numerical data are taken from official sources—the National Science Foundation, the U.S. Department of Labor, and the U.S. Department of Health, Education and Welfare; and from such authoritative organizations as the Engineering Manpower Commission, the National Education Association, the National Aerospace Education Council, the National Science Teachers Association, the Aerospace Industries Association, the Institute of Aerospace Sciences, and similar groups. The latest current figures are used in each case, and the most comprehensive projections into the future.

None of the sources guarantee their figures to be anything but an estimate, based on the most likely yet not infallible analyses of past and present trends. Space statistics, like all others, are subject to variable interpretations, each somewhat different. In some cases, extrapolations into the newest of New Frontiers are incomplete today or even totally absent. Here the author hopes his own calculations, which he was forced to use, were launched somewhere near the proper orbit of the ever-surprising future.

Outside of projected totals, the rest of the data are known facts of today's science technology field in general, the astronautics (missiles plus space vehicles) field in particular. The prophecies of space things to come are on less certain ground but, despite their sometimes eyebrow-lifting predictions, are far more liable to be too *timid* rather than too brash. It is a plain fact that *every* expert's predictions made in 1957, at the debut of the Space Age, have turned out hopelessly (almost laughably) conservative today, only half a decade later.

But we must try to capture some semblance of that breathtaking era ahead. Today is not important to you who are students now. It is the era 5, 10 and 25 years—in fact, a lifetime—ahead when you will live and write the history of the Astronautics Age.

The academic coverage will include the best colleges, the scholarships available and training for technicians in two-year vocational schools. The positions open in government agencies, in the armed forces, and in private industry will be reviewed. All factors relating to how your education will fit you into the vertical frontier of space will be thoroughly digested.

But this book will also try to answer a prime question all of you must have in your minds—*what will it actually be like when I start my astronautics career?*

Therefore, we will not tell you about the typical—and mythical —aerospace firm. We will tell you about *real* firms. How many scientists does each have? How many engineers and technicians? Exactly what do they do? What slice of the space pie does the company have out of NASA's 10-year commitment to place astronauts on the moon? What are prime and sub-prime contractors?

What small firms today have a solid future ahead of them due to pioneering technology? What major fields will grow the fastest during the next 25 years? How does on-the-job training work? What are the 200,000 firms, small and large, that comprise the "space industry" as a whole? What careers will be tops in 1970?—1980?—and on through your lifetime?

In short, how can *you* make the most of the Space Revolution?

Along with this, America's entire space program for some 10 years ahead will be spelled out in the ABCs of precisely what the alphabetical agencies—NASA, DOD, AFSC, all others—are responsible for. The entire stable of space vehicles—from today's Centaur, Gemini and X-15 to tomorrow's Nova, Apollo and Space Plane—will be followed from every laboratory and plant to the launch pad.

Each individual firm's total technical personnel, profits past and present, policy in regard to raises and job advancements, whether it rates high or low to our space leadership—all these are things you have a right to know. If you are being asked to enlist for life in this Battle of Brains, you are obviously entitled to ask what you can expect before you commit yourself.

Space wants you. But first, you want the facts of space. Then you can hitch your mind to a rocket with self-assurance as you set your guidance system to join mankind's drive to the stars.

The author wishes to acknowledge the generous aid of those who supplied statistics, data and illuminating material: Mr. Louis Levine, Director of U. S. Employment Service, *U. S. Department of Labor*; Mr. Austin Stevens, the Director of Public Relations, *Aerospace Industries Association of America, Inc.*; Robert S. Tiemann, Personnel Projects and Recruitment Officer, *NASA*; Richard M. Harbeck, Specialist for Science, Office of Education, *U. S. Department of Health, Education and Welfare*; Mr. Richard T. Fallon, Executive Secretary, *Junior Engineering Technical Society*; Charles W. Webb, Deputy Chief of Staff, Aerospace Education, *Civil Air Patrol*; Mrs. Jane N. Marshall, Editor, *National Aerospace Education Council*; Major James F. Sunderman and Major Gene

Gurny, Book and Magazine Branch, Office of Information, *Department of the U. S. Air Force*; Mr. Roderick Hohl, Public Relations Director, *American Institute of Aeronautics and Astronautics*; also the *National Education Association, National Science Teachers Association, National Science Foundation, American Legion, Science Service, Inc.*, and other public-service and academic institutions too numerous to list; and finally, the public relations and personnel departments of almost every one of the long list of aerospace firms mentioned in the book.

Their cooperation is most gratefully appreciated. However, the use of their data or material does not in any way signify their endorsement of or agreement with the statements, opinions and conclusions reached in this book, which are solely my own.

Otto O. Binder

CONTENTS

GLOSSARY

PART I. *Acronyms and Abbreviations:*

AEC—*Atomic Energy Commission.*
AFB—*Air Force Base.*
AFSC—*Air Force Systems Command* (handling all Air Force space programs).
A-G—*Aerojet-General Corporation.*
AIAA—*American Institute of Aeronautics and Astronautics* (formed by merger of former American Rocket Society and Institute of Aerospace Sciences into one professional society).
AICBM—*Anti-ICBM* (namely, anti-missile missile, Nike-Zeus type).
ALBM—*Air Launched Ballistic Missile* (Skybolt type).
ARENTS—*Advanced Research Environmental Test Satellite.*
ARC—*Atlantic Research Corporation.*
ARPA—*Advanced Research Projects Agency* (of U.S. Department of Defense).
ARS—*American Rocket Society* (see AIAA).
ASSET—*Aerothermodynamic/elastic Structural Systems Environmental Test.*
AT&T—*American Telephone & Telegraph Corporation.*

B.A.—*Bachelor of Arts* (degree; with science major becomes B.S., Bachelor of Science).
BAMBI—*Ballistic Anti Missile Boost Intercept* (future USAF space-defense project).
BMEWS—*Ballistic Missile Early Warning System.*
BOSS—*Bio-astronautics Orbital Space System* (or *Station*) (USAF concept).
B.S.—See B.A.
BTL—*Bell Telephone Laboratories, Inc.*

CY-(1961, 1962, etc.)—*Calendar Year.*

DOD—*Department of Defense* (of the U.S. government).

F-1—*Rocket engine unit* (with thrust of 1,500,000 pounds for advanced Saturn-V booster series).

FY-(1961, 1962, etc.)—*Fiscal Year* (FY-1962 is from July 1, 1961, to June 30, 1962).

GD—*General Dynamics Corporation.*
GD/A—*General Dynamics/Astronautics* (a division of GD).
GE—*General Electric Co.*
GE/MSD—*Missile and Space Division* (of General Electric).
GM (or **GMC**)—*General Motors Corporation.*
GSE—*Ground Support Equipment* (for rockets and missiles).
GSFC—*Goddard Space Flight Center* (of NASA, at Greenbelt, Md.).
GTR (or **GT&R**)—*General Tire & Rubber Co.* (parent corporation of Aerojet-General).

H-1—*Rocket engine* (of Redstone/Jupiter; improved unit used in cluster of eight for Saturn C-1 booster).
HEW—(Department of) *Health, Education and Welfare* (of the U.S. government).

IAS—*Institute of Aerospace Sciences* (see AIAA).
IBM—*International Business Machines Corporation.*
ICBM—*Inter-Continental Ballistic Missile.*
IRBM—*Intermediate Range Ballistic Missile.*
IT&T (or **ITT**)—*International Telephone & Telegraph Corporation.*

J-2—*Rocket engine* (burning hydrogen fuel, with thrust of 200,000 pounds, for upper-stages of Saturn launch vehicle).
JATO—*Jet Assisted Take Off* (see RATO).
JPL—*Jet Propulsion Laboratory* (of NASA, located at California Institute of Technology).

LASER—*Light Amplification by Stimulated Emission of Radiation* (variant of MASER).
LEM—*Lunar Excursion Module* (or the "Bug," two-man unit that will separate from Apollo-C moon-orbiting craft and place U.S. astronauts on lunar surface).
LOX—*Liquid OXygen.*

M-1—*Rocket engine* (burning hydrogen fuel, with 1,400,000 pounds thrust—up-rated from original 1.2 megapounds—for Saturn and Nova upper-stages).
M.A.—*Master of Arts* (degree; with science major becomes M.S., Master of Science).
MASER—*Microwave Amplification by Stimulated Emission of Radiation.*
MHD—*MagnetoHydroDynamics* (alternatively called Hydromagnetics: phenomena in physics of ultra-high-temperature ionized gases).
MHR—*Minneapolis-Honeywell Regulator Co.*

MIDAS—*MIssile Defense Alarm Satellite.*
MIT—*Massachusetts Institute of Technology.*
MMRBM—*Mobile Mid Range Ballistic Missile.*
mph—*miles per hour.*
mps—*miles per second.*
MSFC—*Marshall Space Flight Center* (of NASA at Huntsville, Ala.; Saturn plant).

NAA—*North American Aviation Inc.*
NASA—*National Aeronautics & Space Administration* (civilian space agency in charge of all non-military astronautics projects).
NDEA—*National Defense Education Act.*
NDSL—*National Defense Student Loans.*
NEA—*National Education Association* (of U.S. government).
NERVA—*Nuclear Engine for Rocket Vehicle Applications.*
NSF—*National Science Foundation* (of U.S. government).

PARSECS—*Program for Astronautical Research and Scientific Experiments Concerning Space;* research project of Boeing. A *Parsec* is an astronomical unit of measurement equal to 3.258 light-years, or about 19 trillion miles.
Ph.D.—*Doctor of Philosophy* (degree; with science major becomes Sc.D., Doctor of Science).
PMR—*Pacific Missile Range* (as distinguished from AMR, Atlantic Missile Range, at Cape Canaveral).

R&D—*Research & Development.*
RATO—*Rocket Assisted Take Off* (see JATO).
RCA—*Radio Corporation of America Inc.*
RL-10—*Rocket engine* (burning hydrogen, with 15,000 pounds thrust, for upper stage of Atlas-Centaur, and Saturn boosters).
RMI—*Reaction Motors Inc.* (subsidiary of Thiokol Chemical Corp.).
RP-1—*Refined Petroleum-1* (purified kerosene for rocket engines).
rpm—*revolutions per minute.*

SAINT—*SAtellite INTerception* (former Air Force space defense concept, now superseded by Satellite Inspector Project).
SAMOS—*SAtellite and Missile Observation System.*
SERT—*Space Electric Rocket Test.*
SLAM—*Supersonic Low Altitude Missile* (ICBM concept with nuclear-powered rocket engine).
SNAP—*System for Nuclear Auxiliary Power* (within satellites).
SORTI—*Satellite ORbital Track and Intercept* (project).
SPUR—*Space Power Unit Reactor* (within space vehicles).
STL—*Space Technology Laboratories Inc.* (subsidiary of Thompson-Ramo-Wooldridge Inc.).

TIROS—*Television and Infra-Red Observation Satellite.*

TRW—*Thompson-Ramo-Wooldridge Inc.*

UAC—*United Aircraft Corporation.*
USAF—*United States Air Force.*
USN—*United States Navy.*

X-20—(New official USAF designation of) *Dyna-Soar* spacecraft.
XLR-99 ("X" for Experimental)—*Experimental Liquid Rocket-99* (power-plant of X-15 rocketplane, burning alcohol or ammonia fuel, with 57,000 pounds thrust and throttleable controls).

PART II. *Technical Terms:*

accelerometer—Device measuring all velocity-changes (acceleration or deceleration) of rocket while under power.
astrogation—Literally, "space navigation" (prefix similarly modifies astro-*physics*, astro-*biology*, etc.).
avionics—That branch of electronics that deals primarily with flight vehicles (either jetplanes, missiles or rockets) not reaching space and remaining within the atmosphere.

boilerplate—First full-scale model of spacecraft (such as Apollo), made of boilerplate steel without interior instrumentation, for preliminary tests at *flight-weight* (as distinguished from low-weight mockups of wood or plastic).

cis-lunar—"This side of the moon," i.e. the region of space between earth and moon. Likewise "cis-Mars" and "cis-Venus."
cryogenics—Ultra-low temperatures near absolute zero (− 459.6° F.). Also, liquid oxygen and other gases that liquefy only at very low temperatures are called "cryogenic" liquids (see pyrogenics).

dielectric—Non-conducting material such as glass or wood.
dynamics—"Motion of bodies or action of forces" of any given type, according to prefix: *aero*dynamics (aircraft), *astro*dynamics (spacecraft), *spin-*dynamics (centrifugal force), *gas-* dynamics (high and low pressure).

ecliptic—Path of planets across night sky, resulting from the "pie-plate" plane in which all the planets revolve around the sun.
ecology—Symbiosis (partnership) of plants and animals in mutual support of life (in astronautics, using algae plants to supply oxygen or food to astronauts in spaceships).
electric propulsion (or **drive, engine**)—Any form of rocket thrust derived from high-speed subatomic particles (protons etc.) or charged atoms (ions)

supplied by nuclear reactors; characterized by low but steady thrust, able to propel far greater payloads than chemical rockets. (Electric-drive variants include the Plasma-jet, Arc-Jet, Nuclear Pulse, MHD-propulsion, Ion-drive, etc.).

escape velocity (from earth)—6.94 miles per second or 24,984 mph.

exo-biology—Life possibilities on other planets.

"exotic" (technology, systems or hardware)—The most unorthodox in advanced space concepts.

follow-on—Trade term for next vehicle in a series, or an advanced version undergoing research.

g-(force)—Unit-g pull of gravity and standard weight at earth's surface. During rocket launches, mass increases directly with multi-g stresses, i.e. 10 g's increase astronaut's weight 10 times.

geo—Prefix for "earth," hence geodetics (mapping), geo*physics* (of our world), Geo*Sat* (Geophysical Satellite observing phenomena of earth).

hardware—Generic trade term for all missiles, rockets and space vehicles, or parts thereof, that take form from original "paperware" (blueprints), "software" (formulas) and "ideaware" (concepts).

hypersonic—Beyond supersonic; any velocity of more than 5 times the speed of sound (3300 mph and up).

inboard (or **onboard**)—On or within a space vehicle, such as power-system, telemetry, sensors, any instrumentation.

inertial guidance—Based solely on inertia and momentum forces, independently of gravity, so that accumulative readings are absolute and give accurate distance from launch-point at any given moment, thus resulting in precise missile/rocket guidance.

logistics (lunar, orbital, interplanetary)—System of regular launches, supply craft, crew rotation and all other backup activities necessary to maintain a space station, moon camp or interplanetary outpost.

Mach—Unit of supersonic air-speed, variable from ground-speed depending on altitude and temperature. (Mach 1 is 760 mph at sea-level in 0° temperature, 660 mph at 35,000 feet at −40°.)

megapound—One million pounds, in metric system (similar to radio mega*cycle* and power-unit mega*watt*).

monitor—Any observation device or system.

optimum—Achieving the greatest results from any given astronautics project (adj.: *optimal*; converse: *minimal*).

orbit-transfer—"Change of orbit," in altitude or direction (plane), or both. (All shifts of earth-orbiters would be in upward or downward spirals; interplanetary transfers to orbits of Venus or Mars require solar-orbit trajectories timed to meet target world.)

organo-bionics—"Body engineering" via miniaturized electronic devices to modify metabolic/physiological norms and adapt man to space conditions (lowered food and oxygen needs, deep hibernation on multi-month space trips).

oriented—"Focused, aimed at, slanting toward" (earth-oriented satellite camera always pointing down . . . tracking antenna following distant probe . . . student taking space-type curriculum).

parameter—Engineering term for arbitrary framework of known (but limited) data within which to work ahead on a new project (namely, a set of design assumptions).

plasma—Ultra-thin ionized "gases," as in earth's ionosphere and interplanetary space (often called "fourth state" of matter beyond solids, liquids and gases, because plasma comprises subatomic rather than atomic particles).

posigrade (rockets)—For separation of booster from upper-stages, or manned capsule from launch vehicle (pushing vehicle *ahead* in contrast to *retro*-rockets, which cause slowdown).

prototype—Experimental version of a missile or space vehicle, whose flight tests will lead to improved operational model; also, a basic design for one vehicle that can be adapted to another.

pyrogenics—High-temperature physics, the opposite of sub-zero cryogenics.

rigidized—Coined by space engineers (not in Webster's) for inflatable satellite (Echo type) whose flexible skin is kept firm by a coating of metallic paint (usually aluminum).

rocketsonde—Upper-atmosphere research rocket (non-orbiting) equipped to send back telemetry signals (rather than merely photographic film or other non-transmitted data).

ruggedized—Space Age coinage (à la "rigidized"), meaning space-vehicle instrumentation designed as ultra-rugged and damage-proof.

Saturn (launchers)—High-thrust boosters with "ganged engines" to become operational after 1964. (Note: Just before publication date of this book, NASA gave new designations to the Saturn series: former Saturn C-1 (2 stages) is now Saturn-I; Saturn C-1B (3 stages) is Saturn I-B; and Saturn C-5 is Saturn-V.)

semi-conductor—Semi-metallic elements (germanium, silicon, boron, selenium, etc.) or compounds (certain plastics, ceramics, silicates) that are poor conductors of electricity and whose voltage limits thus become ideal "gates" to regulate circuits with great precision. (This led to famed breakthrough of first germanium *transistor* by BTL.)

sensors—Ray-sensitive devices in any electromagnetic octave (gamma-rays, X-rays, ultra-violet, optical, infra-red, radar, radio); carried in satellites and space vehicles for various purposes.

servo- (mechanism, system, control, etc.)—Widely varied devices all of which respond to human command, via radio or other signals (alternative prefix for same: *slavo-*).

static test (of rocket engines)—Test in which prototypes are fastened within steel frameworks for trial ignitions to test fuel consumption and thrust, and to ferret out "bugs."

super-conductivity—A phenomenon displayed by metals near absolute zero whereby resistance vanishes and electrical current continues endlessly without further renewal.

technology—Applied science aspect of astronautics, both R&D and production of hardware. (Note: All space-technology is *engineering*, not "science." The latter's "hardware" is only standard research instruments or experimental devices, never finished vehicles for actual use.)

test-bed—Space vehicle for testing separate devices or systems that will later make up a whole, usually a new type of satellite or missile.

transducer—Electronic device for converting energy (usually electricity) into transmitted form; also a multi-switch for apportioning electrical power among two or more circuits (sometimes among dozens).

trans-lunar (-Mars, -Venus)—Beyond the moon (or Mars, Venus), as distinguished from "cis-lunar," etc.

zero-g—Weightless state in orbit (but because centrifugal force and earth's gravity-pull neutralize each other, not because gravitation has "ceased").

Careers in Space

1 : THE SPACE BONANZA

(*The growing space budget . . . NASA appropriations . . . Military rocketry . . . The Space Revolution and the brain bottleneck . . . Total existing force of scientific labor in America . . . Mindpower shortage . . . Educational gap . . . Brainpower needs by 1970 . . . Technological womanpower available . . . Soviet women professionals . . . Starting salaries of technical graduates in USA . . . High rewards in astronautics industry.*)

What is the fastest growing field of opportunity in America today for young people to enter?

It is that segment of America's industry called "*astronautics*," or more properly "*missiles and space*"—in short, anything relating to the use of *rocket power*. Whatever it is named, you brand-new citizens in school have a brand-new area in which to sell your talents as engineers, scientists and technicians. It is almost as if this industry were magically created for you. Most of the rocket-age jobs awaiting you did not even exist before you were born. Many of them, in fact, came into being only after the Space Age opened in 1957.

Five years ago, for instance, the national budget of the United States did not even list an item for "Space Research and Technology." The budget for fiscal 1963 (mid-1962 to mid-1963) is another story:

National Defense	$52,690,000,000
Interest on National Debt	9,398,000,000

Agriculture and Support	5,836,000,000
SPACE RESEARCH & TECHNOLOGY	5,774,115,000 *
Veterans' payments	5,298,000,000

* Figure derived from combining civilian program under NASA ($3,674,-115,000) and military space R&D under the Defense Department ($2.1 billion).

Remarkably, in half a decade, space dollars have sprung into *fourth* place in federal funding. Before long, astronautics will undoubtedly move into second place, topped only by military defense. NASA's FY-1964 budget of some $5.75 billion, plus DOD's estimated space military funding of $2.67 billion, will almost accomplish that. Experts foresee this combined figure continuing to climb steeply—$12 to $15 billion in 1965 and $18 to $25 billion by 1970. Astronautics will then be the most enormous industry ever known in human history, far outstripping autos, steel, chemicals and all such lesser giants.

Even in FY-1963, the $5.8 billion listed in the national budget for space did not represent the full amount allocated to rocketry. In the huge defense appropriation of over $50 billion, $10.5 billion goes into military missiles. Routine production (missile manufacturing) takes up most of these funds, but $4.1 billion is earmarked for R&D. Of this, the lion's share is for improvements in the giant IRBM's and ICBM's—Jupiter, Thor, Atlas and Titan—which do extra duty as our present-day space boosters.

Thus, taking the entire sweep of the *rocket and space field*, from earliest research to flying hardware, a grand total of $9.9 billion is somewhat near the true figure for the missile/astronautics industry in fiscal 1963. Therefore, the student today contemplating a technological career in any phase of rocketry, whether for earth or space use, can know that by the time he graduates his field will involve enormous sums of $10 billion and more. Nor is that the end. Next year and each year thereafter the funds will grow larger by multi-billion-dollar steps.

Obviously, both scientific space funding and the space technology portion of the military budget are almost doubling each year.

They will keep mushrooming, in the frank opinion of high NASA and DOD officials. The civilian space agency expects to reach a plateau of $7 billion to $12 billion annually, through the rest of this decade. This, by 1970, would mean a total gross of $70 to $100 billion going into our peacetime space effort, including the great goal of landing men on the moon.

This is considered too conservative by many astro-economists, who foresee an awesome total of $150 billion and over consumed by 1970, due to the mounting pressures of space rivalry with Russia. If, for example, an early lunar landing—in 1965 or 1966—is accomplished by both Soviet and American astronauts, a race to the planets beyond would immediately develop and push space costs to far higher levels than is contemplated today, or even dreamed of.

We have been looking only at space R&D plus rocket hardware. If we take a broader look into all R&D done by American technologists, it will include many non-astronautics industries, such as chemicals, plastics, transportation, cosmetics and down the line. Battelle Memorial Institute, a non-profit organization that tabulates complete USA research and applied science activities, announces that the total all-industry R&D in America amounted to $16.2 billion in 1962, $18 billion in 1963, and might top $25 billion in 1964-65, a major "industry" if it were called such.

If to the astronautics portion of the whole were added unestimated but enormous post-R&D costs—production of non-flying hardware, construction of Cape Canaveral launch pads, growing chains of tracking stations, and all other subsidiary expenditures—the total already places astronautics among the Big Ten of industry. Nobody has yet compiled this set of financial complexities, but some experts freely predict that any year now astronautics will rocket past the king—the automobile industry—into first place. *Permanently.* Motorcars will give place to rockets as mankind's most important vehicle.

It is not a question of *will* that electrifying moment come—but *when.*

To a man, economic specialists agree that we are on the verge of a Technological Revolution beside which the 18th century's vaunted Industrial Revolution compares as a firecracker to a nuclear bomb. It is more often called the "Space Revolution" because, as mentioned in the preface, every phase of pure and applied science known today is in some degree a part of space technology, excepting only the esoteric branches of research such as archeology, ethnology, paleontology and the like.

As the space whirlwind completely overturns industry with mounting fury through this decade and the next, a job revolution will also result. Human hands, as a major tool in modern civilization, will take second place to the human brain. Not just in laboratories but among timecard-punching workers.

This is clearly shown by a milestone the American economy already passed, almost unnoticed. In 1960, out of a total labor force then of 73.5 million, the number of white-collar workers passed the blue-collar group for the first time. The latter includes all unskilled and semi-skilled labor in non-manufacturing fields, plus skilled workmen in industrial plants. The white-collar group ranges from general office help to executives, then into the professions of teaching, law, medicine and such, finally including the fraternity of scientists, engineers and technological technicians.

That switch of the blue-collar and white-collar proportions is continuing in the same direction with a ponderous force that is irreversible. By 1970, the national labor statisticians expect the blue-collar total to drop below 33% of all jobholders, as increasing science technology calls less and less for their obsolete services. As fast as this vanishing breed declines, the demand for people with academic training and science skills will increase—and far faster than America is presently geared to produce them.

This brings us not to a manpower but *brainpower* shortage, already close to critical in the research and technology fields. As of

now, America's total existing force of science-oriented professionals
breaks down as follows:

Teachers (science)		307,000
Engineers		1,001,200
Scientists		422,000
Technicians		869,500
	Total	2,599,700

Teaching (total staff, all schools, 2,276,000) and engineering are
the two largest professions in America. Teachers of science are excluded in the following statistics since they are non-producers of
hardware and are not properly a part of the astronautics "industry." Teaching opportunities for those interested will be reviewed
in a later chapter, however.

To define "technicians"—a relatively new term in the technological era—they are trained specialists (but not necessarily with
college degrees) who aid scientists and engineers in a great variety
of ways. They are the *fastest-growing class* of science technology
personnel and will be fully dealt with in chapters ahead.

Referring to the table above, how many scientists, engineers and
technicians will America need by 1970?

Our total population, based on present rates of increase, is expected to advance from 190 million in 1963-64 to 208 million by
1970. Thus, a proportionate increase in science technology personnel, based on population expansion alone, runs to some 360,000.
To this we must add the yearly increase in science personnel we
now achieve, plus extra brainpower that will be required as a
steeply climbing percentage of the total labor force. Estimates
among authorities vary considerably as to this figure when projected to 1970, but averages around 1.8 million new technical people needed. This would also cover replacements for deaths and re-
tirements that occur meanwhile.

There is no reason to go into the involved details of these statistical computations, which can be reduced to their significant core.
Excluding teachers, an official estimate of what will fill the mindpower gap in this decade is based on the following comparison with

current university graduates in science and engineering, plus the non-degree technicians trained in junior colleges or technical/vocational schools. Round-figure approximations are used rather than exact totals.

	Yearly Number Graduated as of 1963-64	Annual Average Needed to 1970
Scientists*	20,000	28,000
Engineers	35,000	82,000
Technicians†	55,000	150,000
	110,000	260,000

* Includes mathematicians but not medical researchers or biologists.
† Figures unreliable, incomplete in all official sources.

Taking only the degree-winning group (scientists and engineers), we need 110,000 annual graduates, precisely double the 55,000 we presently turn out. How can we possibly gain and train 55,000 *extra* technical minds per year? The answer is not as difficult as it seems —in theory.

Out of the grand total of 515,200 persons who received Bachelor's, Master's and Doctor's Degrees in 1962, those in the liberal arts predominated as usual with 125,000 members, followed by 90,000 in economics/administrative fields. If out of those 215,000 only 25% had shifted curricula, we could get the 55,000 science/engineering graduates needed, above and beyond the 55,000 who did graduate.

But could we expect 55,000 additional men to have the special scientific ability necessary? Another set of revealing figures comes from the Engineering Manpower Commission, which estimates that 17% of the 900,000 yearly male high-school graduates have the aptitude and creative ability for science technology careers. Boys with science aptitude therefore number about 153,000.

Obviously, America's well of youth overflows with considerably more than the 110,000 technical minds we need each year. Unfortunately, not all high-school graduates enroll in college, and of the science-capable group that does, only 25% actually take up science or technology courses. The problem is not where to *find* the human raw product, but how to *motivate* them into becoming scientists or engineers. Over 100,000 first-class minds among American youth are "lost" each year to business offices, salesrooms, advertising agencies, law firms and a hundred other occupations—everything but what they are *most* capable of doing. All this has created the manpower shortage that threatens to soon cripple our space effort.

CALLING ALL GIRLS

We are also losing *womanpower*. The story is even worse in this area. We left the girls out of the above calculations for the simple reason that America's female students are notoriously science-shy, yet have equal mental capacities. About 35% of all college graduates are co-eds—namely some 180,000 in 1962. But less than 20,-000 came out with science-major sheepskins. What inhibits them is the misguided American tradition that women must not invade the man-dominated fields of research, technology, engineering or applied science.

Educators have for years deplored this Stone Age attitude as a waste of brainpower we can ill afford, particularly in competition with Russia.

Soviet figures are significant. Women make up no less than 45% of their professional people. Russian women dominate the field of medicine (75%) and science teaching (70%), and comprise a staggering 29% of engineering personnel. In contrast, America's professional women engineers are a minuscule 3% of the engineering society. The fair sex has been slightly more bold in taking up the sciences, and makes up 5% of USA chemists, 9% of mathematicians, and 10% of biologists. But then feminine physicists slip to a low 2%, and in certain research fields like metallurgy and electronics, the female worker is almost absent.

Some firms have deliberately welcomed women into their laboratories, simply because of the male-mind scarcity. Lockheed, for instance, has 61 women scientists—16% of the staff—in its highly technical Flight Sciences Section. If you are a girl student, listen to this flat statement from Lockheed's hiring chief: "When the emphasis is on brainpower rather than musclepower, there is no 'weaker sex.' "

This is proved by the fact that in competitive high-school technical projects and Science Fairs, girls are winning 40% of the awards, even though they are outnumbered 2 to 1. But only 18% of the eligible college material among teen-age girls ever gets there, and of those that do, only 11% enroll in the physical and applied sciences. *America, in effect, is losing intellectual womanpower at the rate of 50,000 unfilled technical positions per year.*

"We're definitely trying to interest high-school and college girls in engineering," states William Howe, chief recruiter for NASA. "Without interesting women in our space program, we just can't see how we're going to meet our long-range personnel objectives."

There you have it, Miss Student USA. Womanpower added to manpower is the one—and probably *only*—way we can make up that enormous brainpower gap we must fill before 1970.

YOUR PAY PAYOFF

Let us turn to more personal considerations. Certainly, in terms of remuneration alone, none of you school-going Americans of to-day—men or women—could make a more practical choice than the field of science and engineering. According to the Engineering Department of Michigan State University, monthly starting salaries for B.S. engineers have on the average gone up some $25 to $30 each year since 1959. In 1962, the base ranged from $525 to $655 (average $575) as compared to liberal arts B.A.'s at $400 to $587 (average $471). An M.S. got you an average $686 and a doctorate won you a handsome $950 minimum.

Every year each of the big aerospace firms—Rocketdyne, Aerojet-General, Lockheed, Douglas, General Dynamics/Astronautics and dozens more—signs up from 10 to 20 of the best grads *per college.* Your pay starts from the moment you are handed your diploma— sometimes before—and if you and the firm prove to have rapport, raises will come more than once a year.

The average pay rise for all technical people has been 6.5% per year since 1953. The median salary before age 30 is $8750. By 35, if you are among the top 10% of America's brilliant minds, you will hold a responsible job—either in research or administration— at a salary of $15,000 to $25,000 in industry. In government civil service the pay scale is somewhat lower, but this will undoubtedly change soon, since NASA and our other federal space facilities simply cannot afford to forego bidding for the highest talent available.

If Horace Greeley were alive and perceived the great new frontiers awaiting high-powered young minds of today, his advice would be, "Go *up,* young man!" Up into space, mentally, where your red-blooded intellect can meet the keenest challenges and greatest adventures in thought that any generation in human history has had the good fortune to encounter.

2 : WHAT IS ASTRONAUTICS?

(Definition of astronautics . . . Brief history of rocketry . . . Space industry today . . . Types of firms . . . Hardware they produce . . . Prime contractors, sub-contractors, supply firms . . . Astronautics personnel . . . Definitions of Scientist, Applied Scientist, Engineering Scientist, Engineer, Technician.)

Perhaps you, the science student in high school or college, have wondered just *what* a "space career" means, what *kind* of work it will actually involve and in *which* firms and industries.

Aeronautics, as everyone knows, is the science of *air flight*. By substituting the prefix *astro* we get *astronautics*, the science of space flight. In its broadest sense, this means the construction and launch of any kind of vehicle—manned or unmanned—that is hurled up above earth's atmosphere into the void beyond. Some devices are a transitional combination of the two, negotiating both air and space, or *aerospace*. Such is the X-15 rocketplane, designed both for aerodynamic (winged) control and true reaction-type flight into space.

The jet engine and rocket motor are akin in utilizing the principle of *reactive* force, based on Newton's third law of motion—for every action, there is an equal and opposite reaction. But where the jet uses oxygen of the air, the rocket carries its own LOX (liquid oxygen) along. The jet, therefore, is forever imprisoned within the atmosphere under a maximum ceiling (for advanced ramjets) of some 50 miles. Only the rocket motor, with its independent propellant supply, can go on into interplanetary space for limitless distances.

Rockets, of course, have been used in the past for other purposes than propelling machines into space. Discovered by the Chinese as early as the 12th century, small and crude solid-fueled rockets were used for peacetime fireworks and as wartime weapons throughout the centuries before anyone suggested, or even suspected, that they could conquer space.

That thought first came up in the late 19th century. Hans Ganswindt, the "Thomas Edison" of Germany, seems to have first designed a spaceship on paper in 1886, but without any specific data. The earliest scientific writings (aside from fantasy fiction) on space rockets are generally credited to Konstantin Tsiolkovsky of Russia in 1898. However, it was the world's first liquid-fueled missile—flown by Robert H. Goddard of America in 1926—that proved rockets could reach space. Finally, when the famed V-2, based mainly on the brilliant scientific and engineering data of Germany's Hermann Oberth, flamed through the war-torn skies in 1944, it pointed the way toward the stars rather than victory in battles.

Almost by themselves, three human brains—those of Tsiolkovsky, Oberth and Goddard—brought forth the Space Age and launched the mighty astronautics industry of today. It is an apt symbol of the power of positive human thinking in new directions. Their ideaware turned into all the space hardware of the past few years—Sputnik, X-15, Vostok, Mercury craft, Telstar, Gemini and the rest of the shining space-vehicle parade.

ASTRONAUTICS INDUSTRY

Today, the "space industry" comprises some one hundred major aerospace firms plus about 1000 sub-contractors and 200,000 supply companies, with a total labor force of 1,800,000 and an annual payroll of $6.5 billion. America's space program now includes 3 groups of military missile systems (75 separate missiles), 50 spacecraft projects and 17 booster-development operations—none of which existed 10 years ago.

The newness of this industry can be judged by the fact that over 50% of the plant floorspace among the aerospace firms is in buildings less than five years old. Further construction of space-oriented facilities is accelerating at an enormous pace.

In most other industries—textiles, foods, autos, etc.—technical personnel number from 3% to 10% of the total working force. In the astronautics industry, from 25% to 40% are scientists, engineers and trained technicians. In some specialty firms producing "exotic" space ware, the technical help outnumbers any other single group. Most companies in the field have more Ph.D.'s on their payroll than the total number of people in management—not excluding vice-presidents.

The gross funds of $10 billion being pumped into rocketry as of fiscal 1963 (military missiles plus space vehicles) are not all going into the production of hardware. About 30% goes into R&D, including fundamental research by scientists, preliminary engineering plans and mockups, and all testing of prototypes of the unborn final product by technicians.

This broadly splits technical manpower in the astronautics arena into two large groups—R&D and hardware production. R&D is largely dominated by research scientists and design engineers of high calibre. Production is fulfilled mainly by broad-gauge engineers and their teams of technicians.

The astronautics firms can be subdivided into a number of more or less distinct groups devising different products or systems:

Structures. The bodies of rocket vehicles are mainly made by manufacturers (about 50) who formerly concentrated on constructing "airframes" for aircraft but who now have shifted largely toward "spaceframes"—that is, the outer shells of space vehicles, if not what goes inside them. Among these firms are the prime contractors responsible for assembly of an entire vehicle—the Atlas booster of General Dynamics/Astronautics (Convair), the Agena second-stage of Lockheed, the X-15 of North American, and so on. The engines, instrumentation, guidance systems and most other internal parts come from other firms.

Electronics.   This is a very important segment of space technology and *the* fastest growing area of all. Among large firms, this field is typified by divisions within Bendix, RCA, Bell Telephone Labs (AT&T) and General Electric. But a host of small firms that have sprung up like mushrooms are also involved, with unabashedly space-flavored or exotic science names—Pneumodynamics Corp., Semi-Elements Inc., Space Age Materials Corp., Stellardyne Laboratories, Ether Ltd., Astro-Science Corp., Hydra-Space Technology, Electrospace Corp., Astrex Inc. Perhaps the most unusual names are Cubic Corporation and Zero Manufacturing Co. (which in the latter case did not prevent the firm from reporting an above-zero net income 177% higher this year than last).

A vast pool of such electronics firms has swarmed to Highway 128 outside of Boston. Along with the Cape Canaveral "Spaceport," the Houston "Moon Center," the Huntsville "Booster Works" and California's "Aerospace Alley," the Boston-centered "Electronics Row" forms one of the main nerve ganglia of America's space effort.

Rocket Engines.   Makers of these are relatively few in number, since the development of today's space powerplants takes an enormous team of technical manpower plus specialized plant equipment and test-ground facilities. Aerojet-General (Titan), Rocketdyne (Atlas, Thor, Saturn), Thiokol (X-15's XLR-99) and a half dozen others dominate the production of rocket motors using liquid fuels. Such firms as United Technology Corp., Hercules Powder Co. and Grand Central Rocket Co. are strong in solid-fuel technology aimed at developing giant boosters of that type for the near future.

Instrumentation.   This includes a wide variety of guidance devices, electrical relays, miniature computers, sensors, telemetry and radio transmitters, solar-cell systems, satellite batteries and a hundred other indispensable mechanisms for making space vehicles tick.

Rather than manufacturing the devices, designing newer and better ones is the important thing. This is carried on by such university-based outfits as JPL (Jet Propulsion Labs), MIT's Lincoln Laboratory and NASA's Ames Research Center, as well as specialized divisions of big aerospace firms like GE, Westinghouse, Ford and G.M.

Fuels. Liquid fuels such as RP-1 (kerosene), hydrazine and alcohol are produced in quantity by oil companies (Standard Oil, Gulf, etc.) and chemical plants (Dow, for instance). The companion oxidant (LOX) is produced by firms long in the cryogenic (low-temperature) business—Linde (Union Carbide Corp.) and others.

But when it comes to solid fuels, the true Space Age firms enter the picture, since the chemicals and motor are always produced as a unit, and each company mixes its own propellant formula. Thiokol, Aerojet-General, Atlantic Research and quite a few others are active in this field, not only devising space-booster engines but also the half a hundred drive units used in military missiles, which are almost exclusively propelled by reliable solid fuels.

GSE (Group Support Equipment). A rocket is a grounded bird without a launch pad. A substantial share of the space budget goes into what engineers call "the care and feeding of missiles," including the massive gantry equipment required at Cape Canaveral and PMR (Pacific Missile Range), hangars for storage of spacecraft, static test-stands and so on.

A large portion of the total military rocketry funding goes into underground silos for ICBM's, mobile launchers for guided missiles, and other non-space projects of that type. Most aerospace firms take these Department of Defense (DOD) contracts for earthbound rocket systems geared to national defense, as well as the hardware marked for space use.

There are many other sub-fields of lesser stature in the intricate web of work the Space Age has spun for industry: tracking systems, gyrostabilizers, miniaturized components, rocketshell metals, radar and TV monitor systems, blockhouse installations, Space Age machine tools, and the like.

Then, after the main contractor firms mentioned above, come about 1000 sub-contracting companies which produce or assemble a bewildering variety of complex gadgetry whose separate parts in turn are ordered from among the 200,000 supplier firms. The latter are characterized by usually producing only one or a few special items by the thousands, be it transistors, explosive bolts, silicon wafers, "squibs" to start rocket motors, spacesuit helmets, zero-g squeeze-tube foods or the myriad other small but vital parts that make up the 200,000 to 1,000,000 separate items that go into the complete hardware for a manned spaceflight mission, the launch of a moon probe, or the orbiting of a new satellite.

It is this multiplicity of mechanisms which are conceived, designed and produced by the army of 377,000 scientists, engineers and technicians directly connected with the astronautics industry. This brainpower is subdivided into as many niches as the flying hardware they produce.

BRAIN BANK

If you are a high-school or college student who wants to become part of the great astronautics endeavor of America, how can you decide whether to be a scientist, an engineer or a technician? What precisely does each do? What part do they play in producing idea-ware (original concepts), software (paper plans) and the final hardware?

The following broad categories are necessarily oversimplified. They often overlap in this modern age of complexly intertwined technology pursuits. But they will serve to give you some foundation for a basic choice of your career in astronautics.

Scientist. In the classical sense, a scientist presumably does only "pure" research, meaning he has no "project" or even "goal" when he starts out. He is after knowledge for its own sake and seeks its prizes wherever they may lure him. Most of the researchers left in this pristine category are in university laboratories or in the sanctums of non-profit and government-funded foundations.

Comparatively few in number—90,000 out of the national total of 543,000 scientists (including teachers)—they are nevertheless the precious core of mindpower from which all modern technological progress eventually stems. Among these researchers are the potential Einsteins who generate whole new industries with a single virgin concept—such as Madame Curie with radioactivity (electronics), Faraday with electromagnetic phenomena (electricity), Marconi with radiowaves (radio/TV), and Einstein with his mass/energy conversion formula (atomic energy).

In theory, astronautics can be traced back to Isaac Newton and his law of reaction forces, the pure-science foundation upon which rests the whole of rocketry. This pioneer work by geniuses is never finished. Present-day idea-forgers are producing principles that will reverberate powerfully through future technology. Their brain-children have already led to the first experimental MASER and its even mightier LASER cousin, to computers daily evolving closer to an intelligent species of machine brains, to eternal electric currents in the cryogenic (low-temperature) world, to tabulations of deep-space radiations which will save astronaut lives. Most awesome are the cosmological concepts of astronomers, astrophysicists and exobiologists who have peopled the universe with a billion other earths and are hard at work at ways to contact them via radio or some exotic means of space communication.

The scientists who delve into the secrets of the universe, at no beck and call but their own curiosity, keep replenishing the well from which we satisfy our technological thirsts of today and tomorrow.

Applied Scientist. This class makes up the largest group of scientists today (275,000 in all industrial categories) and is growing rapidly to meet the demands of the Space Age. About 75% are in private industry, where they are often given a specific task to achieve, although they still retain freedom in choosing the research route to follow.

Investigating new types of propulsion—nuclear-rocket, ion-drive, plasma-engine, electric (subatomic particle)-reaction—is typical of the areas where applied scientists follow brilliant inspirations which can be readily engineered into reality. Among notables in the applied science area are Dr. Hubertus Strughold of space-medicine fame, Dr. Wernher von Braun in advanced rocket boosters, Dr. Ernst Stuhlinger in ion-drive propulson, and Dr. John R. Pierce in worldwide radio/TV satellites (Echo and Telstar).

In actual fact, a good deal of this "applied" research means applying busy brains to sorties into scientific frontiers as far ahead as any the "pure" scientist reaches. Necessity (hardware) is often the mother of invention (breakthrough theory) in the space field. Applied scientists in astronautics today are uncovering more basic principles in ten years than all the ivory-tower researchers of pre-Space Age times did in a lifetime.

Engineering Scientist. This is a brand-new category today, forged out of the demand for space research tools of incredible ingenuity and fantastic performance. "Fiction yesterday, fact today" sums up this astronautics need, with one added phrase—"obsolete tomorrow." In order to keep ahead of leapfrogging space technology, an engineer may decide to gain his doctorate or, conversely, a scientist may take up advanced engineering. Then, armed with both Ph.D. training and design know-how, the ideaware plus software that leads to hardware can together be produced by the same engineering scientist.

Krafft Ehricke (Director, Advanced Space Systems for GD/A), is typical of this new breed of brilliant scientist/engineer, equally at home in the quiet lab or the bustling design room.

Engineer. Traditionally the designer who fashions scientific dreamware into workable wakingware, the engineer of old has sired an immense family of new offspring whose jobs were actually created by the Space Age, with more coming up daily. The main engineering groups of yesteryear—mechanical, civil, electrical, aeronautical, automotive, nuclear—have branched out or evolved into new specialties, such as rocket propulsion, payload ratios, fuel flow, telemetry, data-processing, materials-testing, g-stress factors and reliability of rocket systems. Brand-new and fathered solely by astronautics is *human engineering*: fitting man and machine together like a hand and glove for the mutual invasion of hostile space.

Spacecraft of any kind are the most intricate machines ever conceived by the human mind. They must be made with far more care and under infinitely finer tolerances than cars, planes, ships or any other earthbound vehicles. Rockets are launched only once (at least as of today) and their first chance is inexorably their last. Their payload instrumentation must operate without any attention from maintenance or repair crews. Their outer shells and inner sensor mazes must survive the brutal conditions of zero-g, raw space radiations and heat/cold extremes unknown on gentle earth.

Only the new breed of aerospace engineers can produce these rocket wonders. The mechanical miracles they routinely perform would have been pronounced absolutely "impossible" by the engineering authorities of a decade ago. It is a startling yet documented fact that in 1953 the ocean-spanning rocket (ICBM) was considered "unattainable" by the majority of scientists. As for satellites and spaceflight, a nationwide poll of missile experts of that time produced the timid prediction that such a giant step into space could not come "before 1975"—at the *earliest*.

What astronautics feats will *actually* be achieved by 1975? They are unknown marvels with which you, today's engineering grads-to-be, will astound the world, even more than yourself.

Technician. This is a category of skilled technical help hardly known more than a dozen years ago. They do not necessarily hold a degree but usually have from one to three years of college study, or the equivalent thereof in a vocational school. Some later obtain degrees while on the job by going to night school, thereby enhancing their chances of advancement and higher pay.

Technicians comprise the most diverse series of workers in the country, more so than scientists and engineers combined. These technical people have but one function in common: serving as aides to the academic aces and relieving them of follow-up chores which merely detain them from going on to the next big job.

All authorities—U. S. Department of Labor, National Science Foundation, U. S. Office of Education—agree that technicians constitute the *fastest*-growing labor class in America today. Like a champion boxer's retinue, or a movie producer's trained staff, technicians collecting around engineers and scientists are becoming indispensable.

Technicians are explosively spreading across the board into every area of science technology—handling X-ray machines, checking out launch rockets during the countdown, attending telemetry monitors, installing satellite sensors, testing missile components, running computer complexes and a thousand other delicate tasks. Their work is sometimes monotonous but never unimportant.

Technicians are not to be confused with those in the blue-collar "skilled trades," such as lathe operators or tool-and-die makers. Technicians must have a basic knowledge of science and display a mental aptitude which would be a waste of intellectual talent in a machine shop. They are, in effect, "backup" scientists and engineers.

In salary, job security and future status they are fast approaching the level of the average engineer. Take heart, if you are a college student who has the sinking feeling you won't make the grade as a scientist or engineer and will therefore be "useless" in the space field. By taking the proper courses suited to your abilities, you may very well become a topnotch technician, more valuable in your job than some of the degree-holders for whom you work.

The academic "in-between" unable to master advanced math, the "good-with-his-hands" type, the shrewd "detail-doer," the "gadgets-are-fun" personality—these are the ideal technicians to staff space laboratories. Starting from scratch a decade ago, the estimated total of some 870,000 technicians in American scientific industries (all types) today is fast creeping up on the army of engineers (over one million) built up for generations back, and will double or triple their number in time. Each technical aide increases the scientist's or engineer's over-all creative effectiveness from 25% to 50%, by freeing him from routine "mop-up" duties after launching a project.

"Routine" is a relative word here, meaning anything that isn't strictly creative on the far-out frontiers of space science. The jobs turned over to technicians change as constantly as the rapid pace of technology itself, with exciting new challenges always ahead. The bulk of technicians are not in production of space hardware but in R&D, very often taking an active part in brilliant breakthroughs.

A better term for technicians would perhaps be "apprentice" engineers and scientists, and many of them are now gaining full status simply through experience and phenomenal service. It is estimated that up to 27% of personnel doing full-fledged engineering work "rose from the ranks," out of the pool of non-degree technicians. In the hard-pushing field of astronautics, it is deeds that count, not a piece of paper labeling someone an "engineer."

The red carpet is unrolled for technicians in such fields as rocket assembly, electronics, space-medicine research, chemical technology, cryogenics and astrophysics. But the welcome mat is also out in 75 scientific, 225 engineering and 480 science technology specialties. Becoming a space technician is almost like being born— you are accepted instantly into the astronautics family with great joy. If you can't be a Truax, a Teller, or a Tsiolkovsky, be a technician. They need you almost as much.

3 : TEEN TECHNOLOGISTS

(Choosing a career . . . What qualities make a scientist or engineer . . . High-school drop-outs . . . How they blight their future . . . Aids to science students . . . Science Fairs, Space Workshops . . . Science Talent Search . . . More education after high school important . . . Junior colleges . . . Vocational schools . . . Technician Training.)

We have gone over, in broad terms, just what the astronautics industry is and how technically trained minds fit in as scientists, engineers and technicians. Now we come to an important question —are *you* suited for a space career? How can you tell—while in high school or early college—if you have the necessary mental talents? The personal qualifications? The right motivations? Will a post in the space front-lines give you satisfaction—or chain you to the wrong job for life?

Many educators and space authorities, dedicated to finding all possible mindpower available in America, have devised various guideposts to help the student of today in determining if he can be the scientist of tomorrow. Almost all of their rules place one key human trait above all others:

—Are you *inquisitive* by nature? Do you want to know the *why* of things?

High value is placed on this one factor because quite obviously the scientist or researcher has an oversized bump of curiosity about all things in heaven and on earth. He will with single-minded zest

spend his life happily delving into the unknown. Wanting to uncover "space secrets" is what impelled Goddard to invent the
rocket.

Other personality clues are also given an important rating:

—Are you *imaginative?* (Designing ingenious devices for space
vehicles, for instance, takes high-thrust imagination.)

—Is your mind naturally *creative?* That is, do you tend to want
to write, compose music, invent little gadgets or do anything that
is *original?* (Even the most mediocre scientific mind, almost without effort, conceives new devices or explores unorthodox concepts
that are simply impossible to the non-creative mind.)

—Are you *observant?* That is, do you quickly see details or grasp
the hidden workings of things in general, whether dealing with intangible theories or visible gadgetry? (Just from seeing a few diagrams of captured V-2 rockets fired in America after the war, Karel
Bossart as early as 1947 visualized a complete new rocket system
for the Atlas ICBM, whose development began in 1954.)

—Can you grasp *abstract* ideas? Can you, for example, see how
time is a "fourth dimension" though it does not enter into the
physical form of any object? (Scientific minds seem to universally
possess this peculiar ability to use "applied imagination," typified
by atomic physicists who deal with sub-atomic particles whose precise shape or form may never be seen by the human eye.)

—Do you have a large *vocabulary?* (The far-reaching mind needs
and utilizes far-reaching phrases and often has twice the number of
words at its disposal as the non-scientific mentality. This word-
gathering attribute usually comes out even in early school years).

—Are you generally *objective* about life, rather than subjective?
That is, do you tend to use *reason* more than emotion in any given
situation? (The axiom of the science discipline is to use *facts* only
in research, and this also carries over into daily life for those whose
intellects try to get behind surface illusions and misconceptions to
the real truth.)

—Do you find school science subjects *fun?* (This is not a trivial
point. Most scientists, when they let their dignity down, will tell

you with small-boy eagerness in their eyes that research isn't work but "the greatest fun on earth." That is why scientists on the average put in the *longest* work week of any people in America—*voluntarily*. In fact, it would take a team of horses—or nuclear rockets —to tear them away from a fascinating new path of research.)

—Do you read many *nonfiction* books? That is, outside of school assignments, and ones dealing with science, inventions, philosophy, the arts or any cultural subjects related to the progress of civilization? (Almost without exception, scientists and engineers in their youth literally soaked their minds with the "story of human thought," finding it more dramatic than the best fiction by the world's greatest literary lights.)

—Are you *persistent*, even to the point of sheer stubbornness? (Edison's famed words—"Invention and discovery are 1% inspiration, 99% perspiration"—are still a truism in modern technology. To make a breakthrough into the unknown, each researcher must often batter down seemingly mountainous barriers before reaching his coveted goal.)

—Are you willing to *work hard?* (To reap those few golden moments of triumph upon reaching a new milestone, the investigator must slog through long hours of plain scientific drudgery, often overworking his tired brain more than athletes ever tax their muscles.)

The above are all only *generalizations* that will not necessarily fit every potential scientist/engineer/technician. If you exhibit the majority of these traits to some degree—or exhibit all of them at various times—you are likely to qualify.

Perhaps one other characteristic, if displayed from boyhood on, is an important item among the mysterious ingredients that tag a creative mind in the bud:

—Have you always had *diversified* interests? That is, have you had several different hobbies or pursued a series of "first loves," such as music, books, making airplane models, collecting postage stamps, studying insects, going bird-watching, making small telescopes, gathering rock specimens or any of a hundred such activities? The keynote is having followed a *variety* of stimulations. (Leonardo da Vinci was equally good as an artist, author, scientific observer and inventor. Aptitude tests indicate that most modern scientists could have made a success of several other professional careers, if they had so chosen, which seems to mean that a sort of "universal creativity" can be tapped by the gifted and need not necessarily be turned to science.)

IS ENGINEERING FOR YOU?

Most evaluators make one clear distinction between scientists and engineers when it comes to *academic* aptitudes. Some renowned scientists—biologists, for example—were poor in mathematics during school years. The chemist need not be an "A" student in physics, or vice versa. Being scholastically good in only one-half of all scientific subjects seems no handicap to a scientist, so long as he knows *one* discipline thoroughly plus closely allied ones that he has studied in depth.

The engineer, however, *must* absorb the complete field of *math* and the *basics* of almost every *physical science*—or he'll never engineer anything except his exclusion from the engineering profession for life. Other strongly desirable if not vital personal attributes for being an engineer are:

—An aptness for *mental visualization.* That is, the ability to see objects and their details in your "mind's eye." (You can then more readily "see" rocket gadgetry you wish to design.)

—A liking or at least no aversion to making *long records, lists and data.* (Much of engineering is such "paperwork," with a propensity for organizing lucid notes and clear-cut sketches of mechanisms never before seen on earth.)

—A general *love of tinkering* with all sorts of gadgets from boyhood on—repairing jalopies, working with Dad's home workshop tools, making models of airplanes or ships, keeping your bike or mechanical toys in top working order.

However, the reader must not take any of these *secondary* qualifications as final "absolutes" which cannot be violated. *Only the first item—conquering mathematics and basic science—is truly essential.* Without those no engineer could work out the knotty details and variable factors of even a good mouse-trap, let alone the ignition switch of the restartable Agena engine. But being all thumbs when trying to fix a loose doorknob would not directly handicap an engineer, since he turns over actual hardware-making to skilled lab technicians and plant mechanics who follow his designs and data.

As for mental visualization, there must exist a percentage of top-grade space engineers who are constantly surprised at what their finished gadget looks like. We may think that blindness of the "inner eye" will surely handicap the engineer in devising class-A hardware—until we remember that stone-deaf Beethoven never heard the brilliant music he composed in later life.

THE GREAT SQUEEZE

The rest of this chapter is directed specifically toward the high-school student. There are approximately 11.7 million of you in our nation's classrooms today, from freshmen to seniors, and including both boys and girls in almost equal numbers. But out of the 2.9 million who should be seniors, only 2.1 million will graduate.

The 800,000 drop-outs who won't get a high-school diploma represent close to a *mass tragedy*, especially for the 600,000 boys (who make up the bulk of this group). In this technological age, blue-collar jobs which require little formal schooling are steadily shrinking. And for even minimal white-collar jobs almost all firms now ask the job-seeker this question first—"did you *graduate* from high school?" Any young man today who deliberately challenges these factors is as good as cutting his economic throat for life.

High-school drops-outs face these bitter consequences for a lifetime ahead:

—The unemployment rate among the 16- to 19-year-old group of drop-outs is twice the national average, and is 27% for the 14-15 year olds.

—Two-thirds of the drop-outs will be earning less than $55 weekly when they are 21 (based on today's dollar).

—At age 25 the drop-out will be earning $500 less yearly than the high-school graduate in general, $900 less at age 35.

—His peak earnings will average $85 a week at age 40.

—At age 55 he will seldom be earning more than $4000 a year.

—His total "lifetime income" will average $178,000 compared to $243,000 for high-school graduates and $347,000 for college graduates.

Obviously, a high-school diploma is worth $65,000 to any young man through his working life. To drop out after two or three years of schooling is to take an automatic "pay cut" of $1300 each year for the next 50 years—or $25 a week for 2600 weeks from youth to old age.

Next year, the 800,000 boy and girl drop-outs will compete for jobs with a new crop of over 2,000,000 high-school and 600,000 college graduates. Odds of three to one against him. And in case jobs are scarce and the drop-out is not working, he will again compete the following year with another 2.75 million fresh graduates. Each year, if he has not found a secure job, he will find a new horde of better-educated job-seekers coming between him and his next job—most likely a temporary one.

All of this is pretty poor mathematics in trying to earn a living. Some high-school drop-outs may have a valid reason for quitting, but the majority who have no excuse will have dug their own pit and will face a lifetime of job uncertainty, low pay and the frustration of falling behind steadily in a society which values and rewards trained brains the most.

A new agency called Mobilization for Youth Inc. has been formed, sponsored jointly by the government and the Ford Foundation, and with the added cooperation of New York City, which volunteered for the "pilot" (trial) experiment. MFY's aim is to prevent drop-outs from meeting the despair of unemployment and turning into juvenile delinquents, as all too often happens. Among various methods of rehabilitation, one is training in vocational schools, so that, armed with new technological skills, the ex-drop-out can win a job as a technician. This movement will quite likely prove successful and spread through the country, as time goes on.

In conjunction with MFY, New York City's Board of Education has announced a new plan—Operation Return—that invites teen-agers who realize that quitting high school was a mistake to sign up again and finish their four years. They can return to the class-room even after several years of absence, as long as they have time to gain their diploma before the age of 21. The old adage "Better late than never" applies to this situation. By going back to school, the drop-out will gain the commodity most in demand—a trained mind.

All over America movements of this sort are springing up, all aimed at guiding youth into better education and better jobs. If you are among the recent drop-outs and have no job, do yourself a big favor and apply to your high school for readmission. Without that diploma you may be gainfully employed less than half of your entire future lifetime. If you want steady pay-checks for the next 25 or more years, you must prepare yourself for the new kind of science-transformed world ahead, instead of just leaving your career up to chance.

As the Kiplinger editors put it in 1975, a book previewing the future of America: "The days when young men and women almost haphazardly planned for their futures have ended." By 1975 *every* human mind will be trained and used to the hilt. Those who pre-viously—*now*, that is—failed to finish high school will in 1975 be like someone who didn't finish kindergarten.

THE SPACE ELITE

There are many aids waiting for young people seeking their careers in all the walks of modern technology and industry. For those who are space-minded and enter that special field, the doors truly spring wide open. The National Science Youth Program (NSYP) was launched years ago by Science Service Inc., a non-profit organization. Its activities are as follows:

—*Science Fair.*   Science groups in nationwide high schools have been formed to aid students in devising individual science projects for exhibit. Each year a series of *Local Science Fairs* is conducted to choose the best projects in each state, which then become eligible for the *Science Fair International.* Honors and awards earned by the winners—both at the local and national level—range from cash prizes to college scholarships, plus a public display of their blue-ribbon science project.

Lately, such projects as space-vehicle designs, rocket-fuel researches, aerospace-medicine studies and the like have been capturing more and more of the main awards. The Science Fair winner with an astronautics project is almost assured a fully paid college scholarship and a lifetime job in our space program.

—*Science Clubs of America.*   More than 25,000 science clubs in night schools, or organized among out-of-school teen-agers and drop-outs, also conduct science projects under the NSYP, plus additional sponsorship from many aerospace firms and private educational groups seeking to build up America's "brain bank." Here again, those who show particular aptitude and brilliance in science research or technology become the "fair-haired" recipients of lavish encouragement and monetary backing to continue study in college. The SCA's aim is to rescue the "forgotten young scientists" and give them an assured niche in the scientific/engineering arena of America's future.

—*Science Talent Search.*   Jointly sponsored along with NSYP by the Westinghouse Electric Company, this program separately awards its own series of *Westinghouse Science Scholarships*, with values from $3000 to a grand prize of $7500. The majority of winners in previous years are now university professors doing major research (from a follow-up survey), while the rest hold important applied-science or engineering posts in private industry.

Almost without exception, any winner of NSYP's various youth-science competitions becomes an eminent member of the scientific fraternity of America, some gaining quick fame for their genius in space-frontier breakthroughs. Even the runners-up—by the thousands—are so benefited and advanced in science techniques that technician careers await them with open arms. As the NSYP's slogan states, any "Youth of Today with a creative mind can become the Scientist of Tomorrow."

—*Space Workshops.* This is another unique program that has sprung up since the Year One S.A. (Space Age), sponsored by NASA, the USAF Reserves, the Civil Air Patrol and the local school board where the workshop is located.

The *Bergen County Aerospace Workshop* in New Jersey is one of the pioneers and has had such success in stimulating young minds toward understanding space science that its staff was invited to tour the country and help organize Space Age workshops in other metropolitan areas. Rocketry, space satellites, aerospace medicine, America's full space program—such topics as these are covered in lectures by space VIP's, plus basic experimentations by the young people in a well-equipped "space lab."

—*Project Minerva.* This space-age name was chosen by the American Rocket Society* for a movement first launched in California, in conjunction with the Los Angeles high-school system. Named after the ancient goddess of learning, it was the first lecture course in astronautics ever offered to teen-agers. It is devised largely as a substitute for the amateur rocketry movement that has gripped young America.

The ARS is totally opposed to youngsters forming their own rocket clubs and producing homemade missiles, citing the injuries and even deaths that have occurred. Though most authorities in America have come out against amateur rocketry, many thousands of such clubs exist throughout the 50 states. Membership runs to an estimated total of 100,000 or more. In spite of opposition, they seem to flourish.

---

* Now the American Institute of Aeronautics and Astronautics (AIAA), after merger with the Institute of Aerospace Sciences.

It is difficult to take sides on the controversial matter of amateur rocketry. Nobody denies that the eager scientific spirit behind their endeavors is far preferable to juvenile delinquency. At the same time, as the ARS puts it: "America needs rocketmen, but *alive*— not dead." Perhaps on that keynote, the thoughtful teen-ager should question whether he is merely seeking "thrills" in the guise of "learning rocketry." He will learn just as much if not more rocketry—without danger—through the Space Workshops, Science Fairs and similar organizations.

To any of you teen-agers with "rocket fuel in his veins," we might point out that by training to become a space scientist or engineer, you can eventually launch the *real thing*—the giant, roaring rockets of our space program. Keep that goal in mind. Shooting off a piddling rocket today—that goes wrong and explodes—may keep you from ever shooting off a mighty Titan or Saturn in the future.

AFTER HIGH SCHOOL—WHAT?

At present, 3 out of every 5 high-school graduates go on to college (70% of the boys, 47% of the girls). What happens to the other 850,000? Most of them stop their schooling and start working. But they too, although they are far better off than the non-diploma drop-outs, are finding the going increasingly tough in the onrushing Space Age, which is part of the greater Science Age into which the world is advancing at a furious rate.

Tackling the long-term job market while armed with a high-school diploma alone, in many cases, will also be a losing game all through life. Even if the graduate does not want to obtain a 4-year college degree, for some good reason, he is urged by educators and other authorities to consider further training in a 2-year junior college, or in some kind of vocational school. It is this further training that can lead those with latent abilities into the wide-open field of *technicians*, the aides to scientists and engineers mentioned previously.

Junior colleges, almost unknown a decade ago, are rapidly springing up across the nation, as the various state educational offices
recognize their value in building up the technical forces America
desperately needs. There are well over 1000 such 2-year colleges
today, in all the 50 states, enrolling a quarter of a million students
yearly. They offer curricula either toward a vocation or toward a
college degree, if the student plans to keep on with his formal education.

There are several important advantages of junior colleges as compared to 4-year colleges or universities:

—Lower tuition costs (about half).

—Strategic location within each state so that many pupils can
live at home while attending.

—Single courses available to those having jobs who want to
brush up on any certain subject.

—An aid to young people who must help support their families
with part-time jobs but can still take part-time schooling.

—An ideal haven for the high-school graduate whose marks were
not good enough to satisfy 4-year-college entrance requirements, or who was simply the squeezed-out victim of our over-
crowded campuses.

—By attending junior college (whose courses count toward
4-year college credits) the squeezed-out student gets another
chance to apply for the big college he desires.

—Besides college-credit curricula, a variety of courses are offered
as job aids, or leading to non-degree professions.

Not every teen-ager, of course, is suited for higher learning, and
his true talents may lie in other directions. In the 2-year college, he
can get an *Associate Degree* in business, agriculture, accounting,
merchandising, salesmanship and such time-honored fields.

In the area of private institutions, vocational and technical
schools are also valuable to those who lean toward acquiring specific non-academic skills in various trades—electronics repairmen,

draftsmen, photographers and particularly engineering or science technicians. However, the applicant is advised to choose his vocational school carefully, as there is no national accreditation system. Inquiry with your state Board of Education usually brings you a list of the recognized trade schools of good reputation.

A new phenomenon is the federal-supported vocational school, which came into being through the National Defense Education Act of 1958. Its specific purpose is to aid and encourage non-college young people to fill the "technician gap" in our space program in particular and in our national technology in general.

At first, oddly enough, there was parental resistance, as the NDEA reports: "Many parents feel that their sons or daughters should become engineers or scientists, regardless of their interests or capabilities, because of the prestige factor inherent in such professional occupations."

This called for some "education" of the misguided parents, and the NDEA is now getting its point across that there is a "dignity of employment and the need for workers in occupations not requiring graduation from a collegiate institution. . . ."

This made sense, especially when it was pointed out that technicians are the fastest-growing group of trained help in America today and are being classed more and more along with engineers and scientists as a "scientific profession." Average starting salary, which for the high-school grad is $3900 per year, jumps for the 2-year trained technician to $4600 and goes as high as $7200 in some cases.

After proving their worth, the federal vocational schools caught on fast and money poured in (matched by state funds). This has built some 500 new vocational schools throughout the country to date, with the rate increasing rapidly each year and enrollments skyrocketing. If you have an NDEA vocational school near you, it is well worth investigation—along with private trade schools—to see which suits your needs in planning your future career.

Deciding whether to continue your education after high school in a 4-year college, junior college, vocational school or technical training institute will involve many personal factors and difficult choices. But nothing should stop you from going *somewhere*, if you possibly can. The worst thing is not going anywhere. The penalty may be that you're not "going anywhere" the rest of your life.

4 : TRAINING YOUR BRAIN

(*Comparison of USA and USSR graduates . . . Weaknesses in U.S. educational system . . . Scholarships, fellowships, grants, student loans . . . Where to apply for student aids . . . College costs . . . Rising government aid through NDEA . . . Choosing your college . . . Space Age curricula available.*)

All the movements to promote science among teen-agers have a profoundly serious purpose, as summarized by this statement of Representative Frank Osmers, Jr. (New Jersey) when extolling the Space Workshop program: "Objective evaluation of the world today indicates that America's future—and the very survival of the Free World—may well depend upon an understanding of aerospace power by our people [and] requires a major educational effort among our young people."

And here we come again to a major factor behind the massive nationwide effort to interest you, the teen-age citizen, in space science and technology careers. Quite baldy, it is because the Soviet Union has *already* begun mining all mental nuggets from among its pre-college students to make sure they are not lost to their space laboratories.

The following comparative statistics are revealing:

Total Annual Technical Graduates	USA	USSR
Scientists	20,000	25,000
Engineers	35,000	125,000
Teachers (technical)	35,000	40,000
Total	90,000	190,000

Is this the result of Russia having more colleges than we have, and thus more graduates? Far from it. The stark fact is that America, with a present-day 3 to 2 advantage in college seats and matriculation capacity, is only turning out half as many minds trained for the Age of Technology. And though the Soviets today are behind in total colleges, they don't intend to remain so. Apparently, according to the future projections, they are building colleges as well as brainpower faster than we are. Somewhere around 1970, if we do not hasten our pace, their capacity may equal ours and then their proportion of science technology graduates will rise to 570,000 annually, while ours (at present rates) will only reach 180,000. This overwhelming 3 to 1 ratio will automatically spell utter defeat for America in the technological race ahead.

All this can be changed, of course, if America decides to move its educational system from the technologically backward 20th (Gregorian) century into the First Century of the Space Age. Our college system must be expanded with speed and efficiency —let us say with half the alacrity with which the Big Leagues recently added teams and spanking new baseball stadiums. And in the Science World Series ahead, we must *motivate* our young people into joining the technological team in large numbers.

If America continues to turn out only 10% of technical grads from its colleges, in contrast to Russia's 40%, we can see the inevitable result. At the present rate, Russia will have made a *net* gain over America of nearly 900,000 scientists and engineers working in their professions in 1970.

By more than doubling the scientific brainweight that the USA trains annually, the Soviets are doubling their national payload of technological achievement, both on earth and in space. Yet we have another rival arising in this arena of international economics —the Common Market in Europe. There are signs already that this European multi-nation group is forging ahead of the USA in automation, thereby producing civilization's products faster, cheaper and often better. They also have plans to become an independent space power, with their own boosters and space vehicles. In a few years, America may be vying with Europe as well as Russia for space honors.

But the competition is terrestrial as well as celestial. Our stake in world trade, as well as in visiting other worlds, is in the balance. Upon the outcome will depend our prosperity at home. In this challenging context, the worst policy America can follow is letting first-class minds slip through its academic fingers. Every youngster with science potential who goes another way is a fearful loss to our national stockpile of brainpower. Our present-day academic sins are apparent from these further statistics:

—Only 3 out of 10 high-school graduates ever become college graduates.

—Of this number (some 600,000 annually) only 6.2% earn engineering degrees, a steady decline from the 11% of ten years ago.

—A combined manpower/womanpower pool of at least 150,-000 latent scientists and engineers drains away uselessly each year, and only 55,000 join those professions.

In summary, it is comparable to a family having a comfortable income of $20,000 a year, yet living penuriously on $5000—*and burning the other $15,000 without ever using it.* America can't win the Earth Race, let alone the Space Race, unless many more of its technological teen-agers become creative science citizens. And it can be done via our student-aid programs on a multiplied basis.

SCHOLARSHIP JACKPOT

For those of you who feel you or your parents cannot "afford" to put you through college, the following review of financial help available may open your eyes.

Scholarships (academic). An estimated 325,000 scholarships to be awarded by colleges and universities are available this year, totaling $150,000,000. These are outright *gifts* that the student never has to repay, and are awarded to deserving students according to scholastic record and financial need.

Scholarships (non-academic).   Another $55,000,000 is awarded by non-profit organizations, business firms, private educational organizations and other public-service groups throughout America.

Student Loans.   Over $100,000,000 was loaned to students last year on the most generous low-interest terms to help them complete their higher education. Some of this loan reservoir is administered by colleges, non-profit foundations and business firms, but most of it is on tap with state and federal educational services.

Scholarship Loans.   This is a recent innovation in which an institution awards a scholarship that must be paid back, but not in money, only in part-time services to the donor. A state fund to a student studying for the teaching profession, for example, means that the graduate instructor then fills in as a substitute teacher for a number of years, without interference with his regular schedule. An estimated $15,000,000 went into this relatively new venture last year.

Fellowships.   A minimum of $75,000,000 last year financed 4-year graduate students in college who wanted to continue toward their master's or Ph.D. degree. This is a gift, with no return payment required.

Grants-in-Aid.   Some $20,000,000 a year from colleges is spread among students who agree to work it off in current services for the college itself, on a part-time basis.

This combined student-aid treasure chest totals over $415,000,-000* this year, and the amount increases annually. Today, there is hardly a student in America—assuming he has fairly good high-school grades—who needs to skip going to college for lack of financial means. A hundred organizations wait on all sides of him to furnish the money for his basic expenses, all the way through to his degree in 4 to 8 years.

---

* Authoritative figures on this total are unobtainable from any reliable source. Estimates have varied from under $300,000,000 to as high as $580,000,000. The general average by majority consensus has been used here.

It must be mentioned as a plain matter of fact that high-school and early college students who show a strong desire and superior ability in the physical sciences (including mathematics and applied biology) and engineering are given *special consideration* in all student-aid programs. And the *topmost attention* is given to those who declare openly they are planning careers in *astronautics*.

This is not "politics," "favoritism" or any other unfair governmental bias. It is simply stark *necessity*, in terms of national welfare, in that the most critical and immediate problem facing America is winning the Space Olympics vs. Soviet Russia. Secondly, rather than promoting "overproduction" of technical minds, it will be making up for years of "underproduction" of vital brainpower. Doubling the scientific quota will merely bring us to the "norm" during the current era of technological explosion.

COLLEGE COSTS

Just what do college costs, for four or more years, add up to? How much of them will be offset by scholarships and other financial aids to you, the student?

Educational costs among the 2600-plus colleges and universities in America add up in 2600 different ways. Public institutions— city colleges or state universities, which make up ⅔ of the total —are considerably less expensive than private schools, particularly for the city or state resident who goes to his local school. Tuition in the city/state group runs from $0.00 to $750, in contrast with $150 to $2800 in the private group. However, it is also true that scholarships run higher in private seats of learning than in public ones.

Average costs for tuition are about $250 in public and $850 in private halls of ivy as of today. But remember that tuition rates, which like all else have risen spectacularly in the past decade, will continue to rise. Experts suggest that you calculate at least 8% more each year, so that fees of $500 in 1963 would inflate to $735 or more in five years.

Tuition, however, makes up much less than half the total costs for a student going to college. Tuition is about ¼ of gross expenses at public and ⅓ at private schools. The other costs include books or school necessities and such living costs as clothing, laundry, doctoring and recreation. Commutation costs must be added if you live at home and travel daily to a city/state school nearby. However, this will be far less than dormitory costs for living at private schools.

Throughout the nation's colleges, total costs average out at $1775 per year in public institutions, and $2850 per year in private schools. Living costs are the biggest single item in both cases: $1000 annually in public and $1500 in private colleges.

HOW MUCH DO SCHOLARSHIPS HELP?

Scholarship Aid. First of all, to make it clear, scholarships *never* pay full college costs for any student, no matter how needy or brilliant. Each scholarship award is based on the individual student's circumstances and is the amount *beyond* what his parents and he himself can contribute. Most parents are able to pay some portion of costs, be it 10% or 75%. Almost all students are also able to obtain part-time jobs and make up another percentage of costs, though they are no longer required to "work their way through college" in the old sense of putting in a 90-hour week of combined studies and chores—a regimen that produced too many "unrugged individualists" who quit without a degree.

These percentages of scholarship aid vary considerably from state to state and college to college, ranging anywhere from $10 to $2500 per year. The average is around $500.

Such topnotch "prestige" places as Dartmouth, MIT and Swarthmore tend to present their students with both high expenses and high compensations. The schools can usually afford lush scholarships in that much of their student body (75% average) comes from high-income families whose sons and daughters need not a penny of scholarship aid. In the average institutions, both private and public, less than 60% of the students require no financial help.

As of today (in a rapidly changing picture) it is estimated that 40% of college freshmen require and receive scholarships. The other 60% are not necessarily all from solvent families who foot the whole bill. Many students do not achieve scholarships for lack of good high-school grades, or because they (or their families) are too "proud" to take "charity" and prefer to get a student loan (which must eventually be paid back, instead of being a free gift like a scholarship). Still other students simply find it too much of a chore to apply at a half-dozen or more sources (unfortunately true) for a scholarship. They instead "make do" on some parental funds and their own small nest egg, plus working the whole four years, including summer vacations, school holidays and spare evenings.

However, the "poor" student who truly needs aid should not hesitate to apply for it, so long as he has a good scholastic record. Among the top 25% of high-school students by grades, a high percentage use some degree of student aid, simply because there are few barriers in getting it, if they are from low-income families. Both factors—need and ability—are important. Ability is broken down into several factors when donors decide which students should be awarded scholarships:

—Good grades through high school, placing you in the top 25% of all your fellow graduates. (Some schools only expect you to be in the top 50%, while others narrow it down to the top 10%.)

—Your score on the college board's particular Scholastic Aptitude Test. (Each college makes its own final interpretations of "aptitude" but a score above 600 or 650 usually means you are rated as "college material.")

—When you show evidence (in an interview, plus a record from your high school) of special personal qualities—leadership, well-rounded personality, a hard worker, exceptionally high marks in some or all sciences, etc.

—If you exhibit high "motivation" in wanting to go to college. In other words, you are not being "pushed" into it by insistent parents, or looking forward to a four-year "lark," or seeking "status" for your lifetime through a diploma rather than on your own efforts. Many puzzled applicants who win their scholarships, when they feel they lack any inner "motivation," find out later the magic key was displaying common *curiosity*—the unmistakable hallmark of an embryo scientist.

WHERE TO APPLY FOR SCHOLARSHIPS

There are three main sources for scholarships: (1) Academic institutions; (2) Social service and private organizations; (3) Government agencies (federal, state, local).

1. *Academic Institutions.* You may write directly to any college or university you are interested in for information about its student-aid programs. Your own high-school Student Counselor will give you guidance on what type of letter to write and to whom. Your school library or nearest public library will also have lists and information on nationwide scholarships.

There is also a nationwide service that, though not inclusive of all colleges, lists the available student-aid programs of a great many. For this information write to: College Entrance Examination Board, Princeton University, Princeton, N. J.

2. *Social Service and Private Organizations.* These include a wide variety of groups, each of which can be contacted through a local office in your community or by writing its national headquarters. They include the National Merit Scholarships (Ford Foundation and Carnegie Institute), various labor unions, The American Legion, Knights of Columbus and other businessmen's clubs, women's clubs, PTA's, church sects (both Protestant and Catholic), various engineering and professional societies, and other non-profit organizations.

A great many business firms grant generous scholarships. Though these are usually restricted to children of their employees, they are sometimes available to any deserving applicant with no strings attached. Ford, General Motors, Metropolitan Life and almost all giant corporations have substantial scholarship programs.

Most of the big aerospace firms—Convair, Martin, Boeing, Douglas, Lockheed and many more—have scholarship programs aimed primarily at aiding would-be space scientists and engineers. Besides sons of employees, any young man may win their awards, by committing himself to first work for the donor-firm for a year or two, after which he is free to leave if he wishes. In some cases, the firm supplies the scholarship with no commitment at all except the hope that you will "return the favor" (work for them) when you have your technical degree, but no stranglehold is placed on you if you go to another firm. Just having helped produce *one more* scientist or engineer for the aerospace field in general eases the manpower shortage for each and every such firm.

3. *Government Agencies.* Federal scholarships (as distinguished from fellowships or loans) are granted only to teen-agers entering the service, who may then qualify for academic training in innumerable technical branches of the armed services. The USAF is particularly active in this field and will often give the gifted student a topnotch education, leading to an officership berth in any of its widespread and growing missile or space technology programs. Signing up involves enlistment but only to fulfill the statutory period of service required, after which discharge can be requested, even with a service-financed education in your pocket.

At the state government level, a much larger and purely civilian selection of scholarships is available to chosen high-school students of any particular state. There is an Educational Guidance Service in your own state to which you can apply for scholarship information.

Counties and cities, in certain areas of America, also have their own separate systems of scholarships, about which your high-school Student Counselor will know more.

A WORD OF CAUTION ABOUT SCHOLARSHIPS

It is only fair to add a realistic "warning" to those seeking scholarships. Obtaining one is not always easy. In fact, you should in most cases be prepared for a rather long and often arduous campaign—writing letters to several sources, being interviewed and perhaps rejected once or twice, having papers filled out by your parents regarding income, pledging your willingness to work and earn part of your college education, taking aptitude tests of various kinds, and many other time-consuming details this book could not bring forth.

Discouragement is more likely to strike you than not before you win out. You may become *bewildered* at the "maze" you are in, wondering why it is so "difficult" to get the promised help. You may come to *resent* the strange paradox of hearing how "important" a trained mind is to America's future, while at the same time you have to seemingly move heaven and earth to get into college to train your mind in the first place.

Part of this you must put down to plain "bureaucracy," which at times seems to work at complete cross-purposes. Some of it can be called "the breaks"—good or bad. Most of it—and this is no fault of yours at all—is simply that despite recent progress, *America's educational program is not yet geared to the maximum effort demanded by the "space revolution" and its handmaiden of technological explosion.*

Russia has "tooled up" its educational system for the Battle of Brains with which it has chosen to deliberately challenge America. *Every bright mind* in Soviet schools is already tagged in *grammar* school, after which the child is "pushed" into scientific studies through high school and college—*with all expenses paid in full automatically.* The science-minded student there does not have to apply for any "scholarship." His education *and* living costs, all the way from grade school up to his Ph.D. (if he's qualified), are taken care of without asking. Even his parents are given "compensation" for the loss of additional income while the student is in school instead of working.

One might be envious of this so-called "smooth road to education" for Soviet youth—except for the *penalty*. Upon graduation, the generous—and steely—Communist grip does not let go but holds on for life. As a Russian engineer or scientist, the price you pay for your "free" education is never to be free to choose your own career. They have made that "easy" for Comrade Student too —he obeys orders and works for the state. Period.

And so, despite the subsidy shortcomings of the American educational system, under our democratic process, you do not have to pay the ultimate price of releasing control over your lifelong career. It is your own, to mold and direct as you wish. The financial aid you get eventually, despite difficulties, is not a mortgage on your brain till death.

These may be mere nationalistic platitudes—or eternal verities. Take your pick.

However, this still does not *excuse* the glaring flaws in our creaking educational system as of today. Regardless of how we reject Russian *motives* or *methods* in creating brainpower, we have to face the stark fact that they are *succeeding*. The small scientific platoons we are under-producing will, within ten years, face a vast army of Soviet technologists. That is why we must double or even triple the brain output of our academic system, at any cost, as many nationally known authorities are constantly reiterating. The present total of $400+ million in student aid of all forms seems impressive until we realize it *should* be $750 million to perhaps $1 billion, if we wish to rescue those two out of three young minds lost to science technology each year.

And that is why you teen-agers of today are still finding it an "obstacle course" to obtain scholarships in spite of growing national alarm over the brainpower shortage. When full action will finally follow need (the former is not necessarily triggered by the latter under loose-jointed democracy) no one can predict, but it is quite likely that within 3 to 5 years there will be a Student Aid Explosion. President Kennedy's Aid to Education Bill (killed in large part by the previous Congress but almost certain to be adopted some year soon) has a provision for creating 100,000 new four-year scholarships each year.

One cheerful factor comes out of this, for you who entered college in the meantime. Once you are *in* college, scholarships and other financial aids are less difficult to obtain, assuming you make a decent record the first year. As the Engineers' Council for Professional Development advises technical students: "It is easier to get a scholarship award after a successful year at college . . . The more [scholastically] successful the year, the better the possibilities . . . Therefore, the student should not hesitate to borrow for that first year . . ." This is an "out" from the scholarship endurance contest, an alternate plan that more and more students are following each year—taking a loan instead of waiting for an elusive financial gift.

Under the spur of the missile threat (then ominous) and the Space Race, the U. S. Government in 1958 enacted the National Defense Education Act, whose NDSL office (National Defense Student Loans) makes federal funds available for students. Adequate loans (usually more than scholarships) are available at the lowest interest rates (suspended until graduation) and most convenient pay-back terms (monthly payments after starting work, sometimes postponed until three years later).

From 1959 to 1962, NDSL granted 450,000 student loans, which jumped an estimated 100,000 over the average in 1963 alone. Also, 7000 fellowship loans were awarded for post-grad studies toward higher degrees. The funds (matched equally by federal and state agencies) are practically unlimited so that many thousands of students can use them as their springboard into college.

A survey by the Office of Education in HEW reveals the following NDSL data:

—71% of the student borrowers come from families whose annual income is less than $6000.

—41% are from families that average $4000 yearly.

—For 92%, the student loan was what enabled them to go to college or to continue after having enrolled. Without the loan, an undetermined number of applicants would have thought twice and then decided not to enroll; and many college freshmen would have been unable to continue.

—81% had pre-college savings of less than $250.

—The majority (61%) borrowed less than $250 for their first year. Another 16% needed $500 or less.

—The largest groups felt they could safely borrow up to $2000 in four years (29%) and up to $3000 (24%).

—The average loan for all, from pre-freshmen to seniors, was less than $600 yearly.

—In 1963 an estimated $200-250 million was loaned by NDSL.

Though less rigid and picayune than scholarship rules, the qualifications for student loans still require good high-school grades, proven financial need and a sincere desire to stick it out and achieve a professional career. As previously mentioned, students who have already decided on a science/engineering career—particularly in the astronautics field—find the student-loan doors flying open for them.

In families of a high income bracket (as defined by colleges), is the bright student ineligible for any financial aid at all? Not quite, since there is one student-aid source that is independent of family financial factors. This source is award "contests," open to any applicant purely on his own merits. For full information, write to: National Association of Secondary School Principals, 1201 16th St., N.W., Washington 6, D.C.

CHOOSING YOUR COLLEGE AND SPACE CURRICULUM

As of today, there are over 2600 institutions of higher learning (anything beyond high school) in the United States. This total includes all two-year junior colleges, four-year colleges and the universities giving up to seven years of training toward the highest degrees. Their enrollments run anywhere from less than 100 in small liberal-arts colleges to over 100,000 in giant universities. Yet size has little bearing on which are the best colleges, and the small can often outrank the big in national accreditation.

How do you choose out of this confusing plethora of schools? This is perhaps the most difficult advice of all for anyone to give you. There are "right" and "wrong" colleges for each individual student, depending on many personal factors impossible to list. Your high-school counselor will give you valuable guidance in this area. The obvious misfit circumstance is a would-be engineer going to a college noted for its liberal-arts curricula. This at least we can eliminate in the case of the technical-minded student—he must obviously choose a technical-type institution.

If you have already made up your mind to be an engineer, or if you at least have a leaning toward that pursuit, the list of engineering colleges found in Appendix I can be consulted with profit. Even if you have already enrolled for two years in some other college, you can of course switch if you wish.

Once you have selected your college, the next big question is —what kind of course to take? Again, educational experts have compiled data for you. Perhaps no other field of academics has been so thoroughly researched for you in advance, by the most authoritative agencies in America. The reason is not hard to find —astronautics engineers and space scientists are needed *badly* in America. Thus, all the forces of government, industry and education are striving mightily to make the way easy for the potential space recruit.

Five years ago, no college in the U.S. offered any degree in astronautics or space sciences. Today, a growing list of nationwide schools has devised special "space" curricula that train the student for specific careers in the New Frontier, either in the scientific or engineering fields. The well-known American Rocket Society has prepared a roster of space-oriented academic institutions; you will find it in Appendix I.

The following *typical* curriculum—by courtesy of the IAS (Institute of Aerospace Sciences)—is only a basic guide and can have many variations, but it should be illuminating to the interested student:

A TYPICAL AEROSPACE ENGINEERING CURRICULUM

The first two years of any engineering program are almost always devoted to the basic engineering sciences, mathematics, and English. While the exact order of courses varies from college to college, the content of these first years in most colleges is essentially the following:

FIRST YEAR	SECOND YEAR
English	Chemistry
Mathematics (Analytic Geometry & Calculus)	Mathematics (Calculus & Differential Equations)
Physics	Physics
Drawing & Descriptive Geometry	Engineering Mechanics
Shop Practice	Strength of Materials
History or Modern Language	Fluid Mechanics
Physical Education and/or ROTC	Physical Education and/or ROTC

Beyond the sophomore year the differences between programs at various colleges are greater, and many engineering colleges are now permitting Aeronautical Engineering students to choose between a program devoted primarily to Aerospace Design and one devoted to Aerospace Research and Development. Such alternative curriculums might be the following:

THIRD YEAR

Aero-Design Program	*Aero-Science Program*
Thermodynamics	Thermodynamics
Electrical Engineering	Analytical Mechanics
Applied Aerodynamics	Electricity & Magnetism
Elementary Structural Analysis	Fluid Dynamics
Mechanical Design	Mathematics (Advanced Calculus & Analysis)
Aeronautical Laboratory	Aeronautical Laboratory

FOURTH YEAR

Aero-Design Program	*Aero-Science Program*
Flight-Vehicle Design	Aerodynamics & Gas Dynamics
Structural Analysis	Solid Mechanics
Vibrations	Physics of Materials
Applied Aerodynamics	Aeronautical Propulsion
Aeronautical Propulsion	Analysis of Missile & Space-Vehicle Systems
Materials & Metallurgy	Electronics
Non-technical elective	Original work (undergraduate thesis)

In other colleges it will be found possible to pursue a course in General Engineering with an Aerospace Option. Such a program might differ from those outlined above, as follows:

THIRD YEAR	FOURTH YEAR
Thermodynamics	Aerodynamics & Gas Dynamics
Fluid & Solid Mechanics	Theory of Structures
Electrical Engineering & Electronics	Materials & Metallurgy
Applied Aerodynamics	Aeronautical Laboratory
History, Literature, or Language	Aeronautical Propulsion
Non-technical elective	Non-technical elective

By courtesy of the AIAA (formerly the American Rocket Society), another general curriculum for a "space degree" is offered. The following are a number of courses advised and arranged by Professor Paul E. Sandorff of MIT as the basic technical disciplines in the field of Astronautics and Space Exploration. Many of these subjects are presented only as advanced studies in some specialized field, and require prerequisite courses. For example, physics of the upper atmosphere may be available only to a graduate in meteorology; telemetering and radio guidance only to a graduate in electrical engineering; and so on. This problem is being given consideration, and many students who enter college now may find a well-developed program leading to a degree in Astronautics will be available.

A) FUNDAMENTAL SCIENCES

1. *Mathematics.* Differential and integral calculus; analytic geometry; partial differentiation; differential equations; Laplace transforms; Fourier series; boundary value problems.

2. *Classical Physics.* Newtonian mechanics; statics and hydrostatics; rigid-body dynamics; oscillations and waves; heat and kinetic theory; electricity and magnetism; optics, atomic physics.

3. *Basic Chemistry.* Atomic structures; mass and energy relationships; rate and equilibrium of chemical reactions; ionic chemistry; acid-base systems; crystals and molecules.

B) APPLIED SCIENCES

1. *Statics and Dynamics.* Forces and moments, stability and instability; kinematics of particles; work and energy; impulse and momentum; rigid-body dynamics; linear oscillatory systems; free and forced vibrations, including damping.

2. *Solid Mechanics and Physics of Materials.* Force transmission; stress-strain relations; stress distribution and deformation; beam theory; shear and torsion; buckling and instability; material structure; material failure; plasticity; stress-strain-temperature-time relations; stress concentration, brittle failure and fatigue.

3. *Electrical Science.* Principles of electrical circuits; components and systems for power, control and instrumentation; theory and performance of electrical components, circuits and systems.

4. *Fluid Mechanics and Gasdynamics.* Dynamics and thermodynamics of real perfect fluids; one-dimensional flow; incompressible and compressible flow, viscous and turbulent flow; lift, drag and boundary layer effects.

5. *Thermodynamics.* Thermodynamic laws; properties of liquids, vapors, gases; heat transmission; gas and vapor cycles.

C) PROFESSIONAL ENGINEERING SUBJECTS

1. *Mechanics of Orbits and Trajectories.* Celestial mechanics, orbits and perturbations; spaceflight trajectories; exterior ballistics of rockets; powered-flight trajectories; optimization techniques.

2. *Propulsion for Rockets, Missiles, and Space Vehicles.* Fundamentals of rocket propulsion; solid and liquid chemical rocket engines; rocket components and accessory design and performance; nuclear, plasma, ion and other systems.

3. *Rocket and Space Vehicle Structures and Materials.* Analysis and design of shells and pressure vessels; elastic and plastic response to

dynamic loads; applied thermoelasticity and aerothermoelasticity; basic loads and materials for missiles and space vehicles.

4. *Guidance, Navigation, and Control for Missiles and Spacecraft.* Inertial, celestial, and electronic navigation techniques; missile guidance systems; control system design for missiles and space vehicles.

5. *Vehicle Design.* Rocket vehicle performance; missile and space vehicle design; practice in system design, combining propulsion, mechanics, aerodynamics, structures, guidance and control for missiles, re-entry vehicles, space vehicles and so on.

D) PROFESSIONAL SCIENCES

1. *Communications.* Radar systems; ultra-high frequency techniques; information theory; microwave circuits; electromagnetic radiation theory and antenna design; telemetering; pulse circuit analysis.

2. *Astronomy.* Planetology; study of the sun, asteroids, comets and meteors; astrophysics; physical astronomy.

3. *Physics of the Upper Atmosphere.* Composition, properties, meteorology of the upper atmosphere; tools of upper atmosphere research.

4. *Hypersonic and Rarefied-Gas Dynamics.* Superaerodynamics; free molecule flow; slip flow; piston theory; boundary layer and heat transfer effects for rarefied gases and gases at high temperature; plasma physics; magnetohydrodynamics.

5. *Space Medicine.* Physiological and psychological aspects of environment of upper atmosphere and space; artificial environments and environmental control.

E) RELATED ADVANCED SCIENCE AND ENGINEERING

1. *Advanced Mathematics.* Probability and operations analysis; statistical theory; advanced calculus; variational calculus; field theory; machine analysis and digital techniques.

2. *Aeronautical Sciences.* Three-dimensional flow; wing theory; classical performance, stability and control; aircraft propulsion; aeroelasticity.

3. *Modern Physics, Solid State Physics.* Quantum theory of matter; quantum mechanics; theory of molecular structure; interaction of matter with electric and magnetic fields; insulators, semiconductors, metals and molecular compounds; relativistic mechanics and dynamics.

4. *Nuclear Technology.* Nuclear structure; nuclear reactions; reactor theory and design; shielding reactor control.

5. *Physical Chemistry.* Composition and physical states of matter;

chemical thermodynamics; kinetic theory of gases; properties of solutions; kinetics of chemical reactions; surface and colloid chemistry; atomic and molecular spectra; electronic structure of molecules; theory of combustion; transport phenomena.

6. *Meteorology and Geophysics.* Physical meteorology; planetary atmospheric dynamics; geochemistry; geophysics; geodynamics; geodesy.

7. *Metallurgy and Materials.* Metallurgical science; advanced materials science; physics of strength and plasticity.

8. *Automatic Control and Servomechanism Theory.* Linear representations of physical systems and analysis of their performance; stability criteria, transient response, frequency response; feedback loops; performance functions by Laplace transforms; analytical, graphical and analog methods; advanced servosystem analysis and design.

9. *Structures.* Advanced mathematical theory of elasticity and plasticity; advanced shell theory, theory of shallow shells; large-deflection theory; methods of optimum design; experimental stress analysis.

For those who are able and willing to continue beyond a 4-year B.S. degree to their master's rating or doctorate (Ph.D.), there is a list of the nation's outstanding centers for technical training in the Appendix.

Most of the colleges that have boldly explored the space education frontier are, of course, not easy to get into. Entrance requirements are necessarily stiff. But the student with a good scholastic record, strong motivation toward astro-technology, and willingness to work his brain to the utmost can find no better place to launch himself into a space career for life. All of the institutes featuring Space Age curricula are heavily endowed with scholarships and fellowships through both industry and government, so that the student's financial problems are relatively light.

However, by no means is it only prestigious "name" colleges that can train your brain for space work. Each year, dozens of other schools are instituting Space Age courses of any and all kinds. A small college may have a surprisingly big program in astronautics. For example, Brevard Engineering College in Florida has a faculty of 25 and a student body of a mere 200. Called the "Countdown College" because it is near Cape Canaveral, Brevard fires forth topnotch engineers of astro-electronics, interplanetary propulsion and other phases of space technology.

It is more than likely that within a few years, all but the strictly liberal-arts colleges will offer some form of astronautics curricula in order to feed the insatiable demand for space scientists, engineers and technicians that our nation's invasion of the celestial realm has created. Willy-nilly, America will have to set its academic sights on utilizing 20% or 30% (rather than today's 10%) of its technical brainpower for the Space Revolution ahead.

That is, with *your* cooperation, young America.

5 : YOUR SPACE NICHE

(Broad choice of career areas open to technical graduates . . . NASA's space schedule . . . Earth satellites and space probes definitely planned . . . Armed forces space projects . . . Boosters and launch vehicles of today and tomorrow . . . Long-range space goals to end of 20th century . . . Non-flying hardware in space program . . . Tracking systems . . . GSE . . . Proportions of funding for pure research and R&D . . . High priority need for space-oriented engineers and scientists . . . Best fields of opportunity for technicians.)

As the college weeks and semesters slip by, a certain date looms ahead of you—the day of graduation. That zero hour will plunge you into the astronautics field and you will be some part of America's space program, either as a scientist or an engineer (or a technician, if graduating from a junior college or vocational school).

What sort of research will you be doing, as a scientist? What kind of hardware will you design, as an engineer? What kind of space project will you be part of, as a technician? In short, which human cog will you become in America's space program?

Looking ahead no more than 5 years, we can name the upcoming projects and hardware with some certainty. From 5 to 10 years ahead, the plans are less firm and may change considerably. Beyond 10 years, the space leadership speaks only in vague phrases. Yet there is no doubt about what, in broad terms, the *rest of this century* will be filled with—space exploration in ever-advancing stages and ever-huger fleets of hardware. Manned orbital flights, lunar landings, trips to the planets, new satellites, deep-space probes, space stations, outposts of the solar system—all these will occupy your lifetime and far beyond.

Even the most scornful space-belittler (the breed has not quite died out yet) has never denied one thing—space is not a passing fad.

NASA'S SPACE SCHEDULE

By far the larger part of our nation's space effort, as of today, is planned and executed by the National Aeronautics & Space Administration, first formed in October, 1958. With a fiscal 1963 budget of $3.7 billion, and some $5.75 billion or more for FY-1964, NASA has laid out fairly firm details for the assault against space through this decade.

The program breaks down into two major portions—unmanned vehicles and manned craft. The two actually work in conjunction, beating out a common path into the unknown, but first we will deal only with unmanned vehicles.

UNMANNED VEHICLES

Up till the 5th anniversary of the Space Age (October, 1962), America had successfully launched 90-odd space vehicles, both earth-orbiting satellites and interplanetary probes. The same number or more will be launched in the next three years according to NASA's official schedule:

Year		Earth Satellites	Space Probes
1963		17	5
1964		24	10
1965		28	12
	Total	69	27

Earth Satellite Types

Explorers—vehicles investigating the ionosphere and beyond (launchings each year, indefinitely).

OSO (Orbiting Solar Observatory)—Follow-on of OSO-1 now in orbit, studying solar phenomena (next 3 years).

OGO (Orbiting Geophysical Observatory)—To make various observations of earth's gravity, magnetic field, etc. (1963 on).

EGO (Eccentric-orbit Geophysical Observatory)—Variation of OGO with an extremely ellipitical orbit to investigate earthly phenomena deeper in space (1963-64).

POGO (Polar Orbiting Geophysical Observatory)—Another variation planned for polar orbit instead of OGO's equatorial path (1963-64).

OAO (Orbiting Astronomical Observatory)—"Telescope in space" with lens up to 36 inches in diameter, for visual study of moon, planets and stars (1964).

TIROS (TV & Infra-Red Observation Satellite)—Follow-ons (5 to 8 more) of half-dozen meteorological (weather) satellites already orbited, to study earth's cloud-cover and detect budding hurricanes (for 3 years more).

Nimbus—Advanced weather satellite based on TIROS system, able to tabulate earth's "heat budget" from the sun (1963 on).

Aeros—Operational WeathSat in high orbit, after Nimbus, expected to give highly accurate long-range weather forecasting all over the world (1965 on).

Echo II—Larger version (135 feet in diameter) of Echo I (100 feet), to further test out passive (non-amplified) relay of microwaves (Project Rebound) (1963-64 on).

Telstar—Follow-on of 1962's active radio-relay satellite, perfecting techniques of relaying transatlantic TV, radio, telephony, telegraphy and photosimile signals (for years ahead).

Relay—Active ComSat designed for higher orbit 1000 to 3000 miles high, with similar performance to Telstar (for years ahead).

Syncom—Higher (22,300 miles) active ComSat on 24-hour stationary orbit with 2 or 4 covering world. Planned as final "operational" system, unless NASA finds the lower-orbit Telstar-Relay system to be more effective (1963 on).

Space Probes

IMP (Interplanetary Monitoring Probe)—Deep-space observers of sun, Mercury, Jupiter and outer planets (1963 on).

Ranger-C—Advanced version with high-resolution TV for better pre-impact scenes of moon's surface (1964 on).

Surveyor—Soft-landing (non-crashing) vehicle to relay panoramic TV scenes of moon's surface to earth (1964-65).

Surveyor-B—Non-landing circumlunar version to monitor moon's radiation envelope, examine surface details and choose best landing site for later landing of astronauts (1965-66).

Mariner-R—Follow-on of 1962's Mariner II Venus flyby craft, with superior sensors to gather more data about planet (1963 on).

Mariner-B—Variant for Mars flyby mission at planet's next close approach (late 1964) to scan Martian surface by TV and release air-entry capsule to seek evidence of Martian life. (Same flyby mission plus planet-landing capsule to be used for next Venus approach (1964-65).

SPACE DEFENSE SYSTEMS

The above are only the space vehicles solely controlled by NASA. The Department of Defense has a lesser list of specialized satellites that are allied to national defense and are thus under the control of the armed services. (No figures are released on how many of their vehicles will be launched in the future, but the number may equal or surpass NASA's total if space defense, as some predict, becomes a major program.)

U. S. Air Force

Discoverers—Series of satellites that release re-entry capsules recovered by aircraft, with space-data instrumentation and as testbed for classified instrumentation in future military satellite systems.

Samos (Satellite & Missile Observation System)—"Spy" satellite with high-resolution TV and still cameras for space surveillance of Soviet territory.

Midas (MIssile Defense Alarm System)—ReconSat with infrared sensors to detect lift-off of enemy ICBM's at their home launch-pads.

U. S. Navy

Transit—Navigation-aid satellites giving all-weather guidance to Navy ships around the world, day and night. (Operational system of 1962-63 will be kept up permanently, with burned-out satellites replaced constantly by new vehicles.)

U. S. Army

ANNA (Army/Navy/NASA/Air Force)—Follow-ons of ANNA-2 now in orbit, a geodetic (mapping or "firefly") satellite with flashing lights (formerly under Army only but the other armed services and NASA now have joint participation).

Atomic Energy Commission

Rover—Nuclear-propulsion rocket in R&D stage, under joint control of the AEC and NASA, with test components to be boosted by chemical rockets until the first nuclear-powered rocket is ready for flight, sometime beyond 1965.

ROCKETS AND BOOSTERS

Besides these major space vehicles, NASA shoots a long series of non-orbiting rockets into the upper atmosphere, at the yearly rate of some 200-plus. The Aerobee, Nike and Argo are typical of this group, carrying aloft a variety of sensors and instrumentation for daily weather conditions and data on atmospheric physics in general. Other up-and-down rockets (Trailblazer) make powered dives to test re-entry devices.

All this earth-departing hardware requires earth-based hardware —launch vehicles, missile sites, tracking stations and the like. The solid-fuel Scout of today and the forthcoming Saturn boosters (flights with payloads starting in 1963-64) will be strictly NASA rockets, but at present all other "space trucks" are modified military missiles designed and developed by the USAF. NASA contracts and pays for as many USAF launchers as its budget allows.

MANNED CRAFT AND 40-YEAR SPACE PROGRAM

NASA has just announced its long-range space goals, in more specific terms than heretofore. This includes the rapidly increasing factor of manned spaceflights to come (but also see chapter 9 for further details involving the Apollo Moonflight Program). A timetable of awesome scope is now firmly on the agenda for the rest of the 20th century, committing America to explore most if not all of the solar system:

	Earth Orbit Vehicles	*Lunar Missions*	*Planetary Projects*
Unmanned Vehicles (Now to 1968)	*Operational Working Satellites* World Communications Network Meterological (weather forecasting) Navigation-aid System, earth-wide *Research Satellites* OAO/OSO astronomical laboratories OGO earth-study group *Engineering Research* Test-Beds	*Ranger* probe impact series *Surveyor* circumlunar and moonlanding monitors *Cislunar and translunar* spacedata probes	*Mariner* flybys of Venus and Mars IMP (Interplanetary Monitor Probe) series *Voyager* (capsule landing) vehicles to Mars and Venus *Out-of-Ecliptic* and *Gravitational* experiment probes *Outer Planet* probes to Jupiter and Saturn and their moon systems *Trans-Solar System* probe (between stars)
Manned Spaceflight (developmental phase)	To 1968 1-man *Mercury* 2-man *Gemini* 3-man *Apollo* craft *Orbit-Rendezvous* experiments *Test-bed* orbit laboratory	Before 1970 *Circumlunar* flights and *Manned Landing* with Apollo-B and Apollo-C craft LLS (Lunar Logistics System) surface vehicles for establishing manned base on moon	After 1975 *Mars landing* *Venus recon* flights *Search for life* on Venus and Mars *Planet bases* (tentative non-authorized projects as yet)

Operational Manned Spaceflights	After 1968 Orbital Operations	After 1970 Lunar Station	After 1980 Mars Station
	(satellite servicing and repair)	(permanently manned and supplied)	(manned and maintained)
	Orbiting Labs (scientific experiments)	*Scientific Observatories* (telescopic; radiation; deep-space radio/radar listening posts)	*Mercury Manned Expedition* (circum-planet, or landing if feasible)
	Orbit-Rendezvous Techniques (Gemini and Apollo, probably semimilitary project)		*Jupiter Manned Exploration Mission* (landings on moons only)
	Ferry Vehicles (transfer of personnel earth-to-orbit and return)	*Lunar Exploration* (Earth-facing and Other Side surveyed and mapped)	*Outer Planets Project* (to Saturn's moons, Uranus, Neptune, Pluto, pending ultra-speed nuclear/electric drives)
	Recoverable Boosters (piloted, via ramjet or retro-rocket engines)		

As of today, much of this is under feasibility study only, particularly the missions after 1970. But most of the pre-1970 earth-orbit and lunar programs are in various stages of hardware R&D. Some of the unmanned "service" satellite systems are close to operational (by or before 1965)—NavSats (Transit), ComSats (Echo, Telstar, Relay, Syncom) and WeathSats (Tiros, Nimbus, Aeros).

NON-FLYING HARDWARE

Along with all of this earth-departure hardware goes a massive conglomeration of earth-anchored equipment without which the fleet of space vehicles would be utterly grounded:

GSE Hardware (Ground Support Equipment)

This includes the launching pads, service towers, gantries, hangars and housing at Cape Canaveral, Wallops Island (Scout launchings) and PMR (Pacific Missile Range). The latter, mainly used for military-missile test shots into the Pacific, or for satellites sent into polar orbit, stretches along 120 miles of California coast and has three separate launch centers—Point Mugu (USN), Point Arguello (USN) and Vandenberg Field (USAF).

Tracking Network

This takes in several chains of tracking stations, some extending around the world, which are the electronic eyes and ears that keep tabs on our growing flock of artificial moons, far-straying interplanetary planetoids, the re-entry vehicles of the Discoverer Program, and the dramatic manned spaceflights. This network has several parts:

Minitrack. This comprises nearly a dozen microwave radio stations, extending roughly north and south from the USA down through South America, for specific radio-beacon tracking (nonradar) of unmanned satellites in equatorial (east-west) orbit. Contact lasts only as long as signals come from a satellite's radio beacon.

SPASUR(SPAce SURveillance). Spasur is a U. S. Navy chain of radars spanning the southern USA from coast to coast, forming an "electronic fence" to spy out all polar (north-south) orbiters, American or otherwise, with or without radio/telemetry transmissions.

SPADATS (SPAace Detection And Tracking System). A group of high-powered USAF radar stations whose exact number and locations are classified, Spadats can reputedly spot any object orbiting within 3000 miles of earth. Up to 500 miles it can spot incredibly tiny bits of space debris, such as broken-off spin-weights from satellites or stage-separation ring bolts no bigger than a thumb.

Mercury Network. This is a chain of 18 stations (to be increased to at least 23 for Gemini flights) straddling the world from the West Indies through Africa, the Indian Ocean, Australia and the Pacific, finally crossing the USA from California back to Cape Canaveral. This chain will continue to serve for all manned flights in the Mercury, Gemini and Apollo-A (earth-orbit) Programs. When astronauts are launched into polar orbits at some future time (probably for military training missions) the USN/USAF tracking chains will go into operation. Lunar and interplanetary flights will be tracked by the following network:

Deep-Space Tracking Facility. Three huge dish-receivers are spaced about equidistant around the world, in order to keep in constant touch with moon or planetary craft. As the earth turns and cuts off each one from its line-of-sight contact—Goldstone in California, Johannesburg facility in South Africa, and Australia's Woomera—the next station "rises" above the vehicle's horizon and gets a "fix." The 85-foot steel-webbed dishes at these installations are sensitive enough to receive earth-oriented (focused) radio signals from vehicles 50 to 75 million miles away—beyond Venus or Mars. If necessary, the giant new 300-foot radiotelescope antenna at Green Bank, West Virginia, can be swung to listen for faint telemetry signals that the smaller dishes have lost, and it is probably capable of reception from Pluto's distance of 3.5 billion miles from earth.

Miscellaneous Hardware

Other unglamorous paraphernalia in our total space effort include trucks and barges for plant-to-pad transportation, innumerable buildings and laboratories for research, computers and data-processors, spacesuits and training devices for the astronauts, and all other adjunct apparatus.

SPACE JOBS UNLIMITED

That is the story of space hardware. But there is still another large category of non-space rocketry—military missiles—which takes up the services of many scientists and engineers. Briefly, some 75 current military missile systems, either operational or in R&D, are carried on by the three armed services. The Army's Honest John, Nike-Hercules and Hawk are representative, as are the Navy's Tartar, Sidewinder and Subroc. Again the USAF has the largest program, including the mighty Titan and Atlas ICBM's, Jupiter and

Thor IRBM's, and smaller air-to-air weapons carried by jet craft, both fighters and bombers.

The actual production and handling of all this hardware, of course, is primarily the province of engineers, plus their assistant technicians, and virtually excludes scientists. However, over 30% of all hardware costs are in R&D where creative ingenuity is often needed, calling for the services of applied scientists and the select echelon of engineering scientists.

And hardware, whether R&D or operational, is only part of the *total* space picture. At least 12% of the budget goes to pure re-search in advanced fields. Future projects, such as interplanetary trips, require extensive "feasibility studies" in which the greatest imagination of the creative scientific/engineering mind is required.

In short, you have an almost unlimited choice of work in the astronautics field, either as a scientist or an engineer, and in a hundred different arenas, from ivory-tower research to fashioning assorted hardware.

As the end of your second year of college approaches, you face a major choice. You want to join the space program (assuming you do well in science and mathematics in general), but should you be a *scientist* or an *engineer*?

Many—too many—students switch from an original engineer-ing course to pursuit of a science degree, for various reasons—they are below par in math (vital in engineering but less so in science), or they find the engineering curriculum too "tough" in general (it is unquestionably the most mind-taxing of all curricula), or they are lured away by the "glamour" of being space scientists, or they feel that the science gap is the most important one to fill.

True, there is a shortage of space scientists, but there is an even greater dearth of engineering manpower coming out of our col-leges. Thus, engineers are *most* in demand (except for non-degree technicians), while scientists run second (or third). For the mu-tual benefit of yourself and your country's space effort, you would be well-advised to stick to engineering (if you find yourself so suited in the first place).

The engineering profession has had a long and honored history, as witness this pledge members make when joining one of their societies:

"I am an Engineer. In my profession I take deep pride, but without vainglory. To it I owe solemn obligations that I am eager to fulfill. As an Engineer, I will participate in none but honest enterprise. To him that has engaged my services, as employer or client, I will give the utmost of performance and fidelity. When needed, my skill and knowledge shall be given without reservation for the public good. From (my) special capacity springs the obligation to use it well in the service of humanity, and I accept the challenge that this implies . . . Since the Stone Age, human progress has been conditioned by the genius of my professional forbears . . ."

The last phrase is no more than plain truth. The first wheel came from a primitive "engineer." Archimedes with his water-screw device was an early engineer using intuitive applied-science principles before the age of science began. Such ancient marvels as the Colossus of Rhodes, the Great Wall of China, the Taj Mahal and the Egyptian pyramids are crude masterpieces of engineering that still strike wonder in modern minds.

Leonardo da Vinci, with his sketches of aircraft, submersibles and many other mechanical devices, was an inventive engineer far ahead of existing technology in his day. And when systematic science did blossom forth vigorously in the 18th century, it inspired an army of technologists who turned the new ideaware into hardware miracles and built up modern civilization.

And in our space program, it is engineers or engineering scientists who are making their indelible mark as world-renowned space VIPs—Wernher von Braun as the most outstanding example with the V-2, Redstone, Jupiter-C, Juno and Saturn. Astronautics annals of the future will acclaim the designers of all the astounding space hardware that sends our robot sensors and living astronauts on and on into the infinite universe.

In *every* field of engineering today, there is a *shortage* of recruits. Your employment prospects are supreme.

On the other hand, if your bent is definitely toward the more theoretical practices of a science discipline, by all means take up that profession. The scientific researcher is the fountainhead of all true progress. Serious manpower gaps have developed in certain theoretical areas vital to our space program.

The U. S. Department of Labor finds that in the next decade, the types of scientists most needed will be in this order—mathematicians (for computers, space calculations), physicists (especially in electronics), chemists (in fuel technology), metallurgists (spacecraft metals) and biochemists (space-medicine program). Yet mindpower in all the major physical sciences (chemistry, physics, mathematics, astronomy, electronics, cryogenics *et al.*), earth sciences (geophysics, meteorology, mineralogy) and life sciences (biology, microbiology, bio-astronautics, astrobiology) are desperately needed in the omnivorous maw of "space science." Along with engineers, scientists in our astronautics industry will inevitably gain nationwide prestige and honor for spearheading America's gallop (it is that) into the space frontier.

But it is worth repeating that if for any reason you fail to finish college, you can still join the *fastest* growing group of trained mindpower today in our space effort—the technicians. It is suggested that our present total of some 870,000 technicians throughout the nation must grow to over 2 million by 1970 if America is to make the most of its scientific/engineering brainpower. Five out of 10 technicians become engineering aides, and 3 out of 10 are scientific assistants. As fast as the scientist/engineer group grows, the technician class must grow at a *doubled* rate in order to meet the insatiable demand.

A joint report of the National Science Foundation and the U.S. Department of Labor states that the following fields should be watched for the greatest growth of technical aide personnel in the Space Age: *statisticians* (compiling and organizing massive data on space research and rocket hardware), *draftsmen* (converting engineering plans into blueprints), *electronic mechanics* (for servicing giant computer systems) and *computer programmers* (to "translate" problems into the computer's "language").

Newest of all, the latter group numbers less than 50,000 programmers in the USA today. At the explosive rate with which electronic brains are being geared into our space program, the specialists who feed questions into these mechanized minds will have to quadruple in number, at the very least. Any student with a reasonably good facility in basic math (such as displayed by the average CPA) can take courses and become an employable programmer.

Most of the technicians mentioned above can obtain the required training in two years of a technical junior college or a comparable vocational school. After schooling, pay on the average starts around $4600 a year and can grow in certain fields to $7500 in a relatively short time. Cases that run over $10,000 a year are exceptional—but only today. It is predictable that increasing demand for space technicians, plus manpower shortages for many years ahead, will by 1970 put their average pay scale closer to $10,000 than $5000 a year, with a good percentage going on to the $15,000 bracket.

Among many general employment services, two are particularly tailored to bring man and job together:

> DECISION, INC.
> Technical Recruitment Division
> Cincinnati 27, Ohio

> CAREERS, INC.
> Technical Personnel Placement Service
> Washington, D. C.

Are they eager to hire you freshly graduating scientists, engineers or technicians? David O'Brien, a job broker (who calls himself a "headhunter"), scours the country for technical talent. All expenses are paid by aerospace firms, who give him lists of their most dire manpower needs, which run to 300 a week or more— only a fraction of which he can fill. He revealed that it took $9000 in advertising and interviews to recruit *two engineers* for one firm. A $38,000 marathon manhunt netted just 38 specimens.

They are out hunting for you, or patiently waiting for you to graduate. In the Space Age, *brains pay*.

6 : ALL ABOUT NASA

(Choice of job field: NASA, academic research, armed services missilery, astronautics industry . . . Salary comparisons . . . Employment totals . . . Private industry's dominance . . . NASA careers, facilities, types of space work . . . NASA prestige and civil-service advantages . . . NASA recruitment programs . . . NASA job-training, fellowships, other grants.)

You have made your decision whether to be a scientist or engineer, and will graduate with at least a B.S. degree. You have chosen whether to follow research or applied science. You even know what particular area of the space program you would like to be in.

Now another big question looms—just *whom* will you work for?

Do you sign up with *NASA*, which amounts to civil service or working for the government? With an *academic institution,* which will include both teaching and research? With the *armed forces*— the USAF, Army or Navy—because you are subject to call for duty? Or finally, with some *aerospace firm* in private industry?

Those choices face each of you scientist/engineer grads as you clutch your diploma and march out of the portals of learning into —whose portals? Which will be better for you? Offer the best pay? Incentives? Challenge? Or satisfy whatever individual motivations are within you?

Many of these inner subjective factors no outside person can solve for you. But we can break down for you certain specific facets about each area of employment, which will help you make up your mind.

We will start out with a generalization, which, like all such sweeping statements, is true for the *majority* but not necessarily true for any one *individual*. It must be taken with the proper reservations as to what it means for your entire lifetime ahead, not just tomorrow.

The statement is that private industry in the astronautics field hires the *most* men and pays the *best*.

Some 80% of the space budget is awarded in contracts to businesses. The remaining 20% is divided among NASA technical personnel, those in the missile programs of the armed services, and various colleges and universities doing space research under NASA/DOD sponsorship. Obviously the job opportunities are far greater in private industry.

EMPLOYMENT TOTALS

At present, the total number of existing scientists and engineers in America (excluding teachers) is close to 1,425,000 persons, who work in *all* fields whether astronautics or not, and in all areas—government agencies, academic laboratories and industrial plants. Space itself, rising by leaps and bounds, today takes the greatest single share of mindpower—15%, or 220,000. Of these, about 80% or 175,000 are engineers and 45,000 are scientists.

If you are aiming for *any* industrial area of science/engineering —including such earthly fields as food-processing, automobiles, plastic products and the like—your choice of jobs is, of course, much wider (1,300,000 job holders today) than if you narrow your ambitions down to space work only.

However, the greatest manpower *shortage* exists in the astronautics field and thus the *demand* is greater there. And since our colleges are turning out *less than half* of the new scientists and engineers we need annually for the mushrooming space program, you will find *more* jobs available each year for a number of years ahead, until our education gap is solved.

Your brightest possible prospects are in figuratively putting out your shingle as a space specialist. You can hardly lose, except for one thing—your unemployment collections for life.

The combined quarter of a million technical minds working on space in the hundreds of aerospace firms, as mentioned above, are those holding degrees only. The figures did not include technicians, on whom official estimates are either absent or very incomplete. According to the most informed sources, about 157,000 technicians work in the astronautics field. If "technical aides" are included—people with very little academics beyond high school but a great deal of practical on-the-job experience—the total rises 33,000 more, to a combined figure of 190,000 who serve as assistants to space scientists and engineers. This field of skilled technical employment, as mentioned, is growing at an enormous rate.

SALARY COMPARISONS

The second big point in "favor" of private industry is higher pay than government civil service (NASA and DOD) or academic institutions, as the following table shows (based on *averages* for scientists and engineers of *any* technological type but with B.S. degree *only*):

	Academic	Government	Industry
		(by law)	
Starting Salary	$ 4248	$ 5335	$ 6576
After 5 years	6400	7560	9984
After 10 years	8500	12,210	16,476
Administration or management rank (15 to 25 years)	12,300	15,335 to 21,000	20,000 to 40,000

With a higher degree, pay automatically increases over the B.S. figure in all the above categories, by some 10% for the M.S. and 20% for the Ph.D.

Almost every type of engineer known—excluding agricultural engineers and such—is a vital part of private industry in the space field. But certain of the sciences, by their very nature, have relatively low application to the pragmatic world of business. Those special science fields are as follows:

Astronomers. Only some 500 professional astronomers exist in the U.S., of whom three-fourths teach part-time. The space industry as such has little practical use for cosmic theorists, although their speculations about the planets and the solar system are important for planning interplanetary probes. Only star-gazing specialists such as astro-physicists and experts in celestial mechanics take an active role in our space program.

Biologists. Biology is a specialized field, in that out of a total 100,000 members in America, 90% are in medical research for non-profit institutions. The 10% in industry will increase, however, as more aerospace firms carry out their own bio-astronautics research—life-support systems in spacecraft, ecology of future extraterrestrial colonies, growing plants in zero-g space-platform "farms," and other exobiology avenues that open up constantly. Also, NASA and the USAF, which have recently increased their astronaut and space-pilot rosters, will call for more attendant bio-experts and space medics as time goes on.

Esoterics. Such scientific branches as archeology, paleontology, ethnology and oceanology have little direct application, naturally, to our space effort.

But all the physical sciences and their many branches—chemistry, metallurgy, nuclear physics, solid-state phenomena, electronics, molecular structure, hydromagnetics *et al.*—are extremely vital to our space program, as are *all* engineering fields. Even civil engineers building bridges, roads and canals are now being asked to plan routes over which giant future boosters and space vehicles can be shipped cross-country from factory to launch pad.

Technicians are prime prospects for whom a special astronautics destiny awaits. To give more significant data, the current nationwide total of 869,500 technicians (estimated) in all industry must be sharply increased, according to manpower experts, to at least 2,000,000 by 1970. The less than 200,000 now engaged specifically in astronautics must accomplish the largest degree of increase and strike for a 1970 total of 750,000.

As space technology demands intense brainwork above mere skilled handwork, blue-collar help will phase out. This manpower gap must be filled by "black collar" (for black space) brain skills enabling workers to use Space Age tools—X-ray tubes, interferometers, Geiger counters, radio-isotopes and other scientific devices—instead of lathes and rivet-hammers.

NASA CALLING

We can only briefly point out the space careers open with government civil service, the armed forces and academic institutions (which, to be fully covered, would merit a separate book). After a brief résumé of these fields, the rest of this book will deal with the giant field of private aerospace firms making up the explosively expanding astronautics industry.

Partly inherited from the defunct NACA (National Advisory Committee on Aeronautics), which it superseded, NASA's space empire on earth comprises both various basic research and R&D centers to lay the groundwork for its huge program of invading space and landing men on the moon before 1970, and numerous other administrative and technological installations.

Total personnel of NASA is now approaching 30,000 and may triple before 1970. Some 40% are scientists and engineers. Houston's MSC—already called the "Moon Center"—will be by far the largest NASA facility when completed, since it will handle the entire manned spaceflight program through the Gemini and Apollo Programs to the lunar landing of astronauts—and beyond. In time, some 24,000 people will occupy its various buildings, including thousands who are representatives and liaison experts from aerospace firms closely connected with the spaceflight program. Over 50% of all personnel will be technical men and women, representing the cream of the crop in national brain-power which the manned-flight project is magnetically attracting.

And this brings up a special point—working for NASA will most likely be the epitome of *prestige* in the entire scientific world. Despite having to accept lower pay on the average than in industry, America's brightest young minds are starting to flock to NASA. After all, what can be more stirring than being in the very Mecca of Space and directly partaking in the greatest venture in man's history?

Those of you in college today might well ponder this special NASA aura, which cannot be bestowed by private industry. In its drive to procure the best possible talent since the moon program became official (May, 1961), NASA has lured many a brilliant young brain from college or industry by citing the following advantages:

—A chance is offered to work alongside some of America's greatest scientists and engineering geniuses—Wernher von Braun, William S. Pickering, Robert Jastrow, Hugh L. Dryden, Abe Silverstein and company. It is truly a Who's Who of Space Technology.

—Rapid advancement is almost inevitable, particularly in the next 5 years, what with the fantastic growth of the civilian space agency into America's greatest single pool of brainpower. Researchers in their twenties are already achieving notable breakthroughs and winning coveted science awards. Managers of various projects average 35 in age. It is truly a "young man's game."

—NASA's government-blessed prestige and overflowing wells of publicity guarantee that the spotlight will shine on its individual stars, winning them not only national but international honors, building up to a lifetime reputation as an ace of space science.

—Out of sheer desperation, in the battle with well-heeled private industry for fresh talent, NASA has already won special civil-service concessions—higher ranges of pay for the same time periods, advancements by merit instead of by the book, and cash awards for outstanding science technology feats that aid America in the Space Race.

—Choice of job and place is another "fringe benefit" NASA freely offers new members. NASA has nothing to lose, since there are always openings at any of its facilities throughout the country. NASA is only too overjoyed when you fill *any* position.

Besides this, when objectively evaluated for the long term (a lifetime), civil service does not come out badly versus private industry. Like any post-office clerk, a NASA technical expert is guaranteed long vacations, generous sick-leaves, comprehensive hospitalization plans and retirement at good pay.

Also, unfailing salary increases come yearly for 10 years, and after that each 18 months until the statutory limit is reached. This final limit was raised above civil-service ceilings of $18,000 a year to $21,000, but this is hardly enough, since private firms promptly began bidding for key talent at $25,000 and up.

NASA, in turn, has planned its next move, as reported from the Manned Spaceflight Center in Houston, Texas. The space chiefs are asking Congress to make a special ruling whereby NASA employees are not precisely in civil service, but in a category of their own—at least in the salary scale. Pay raises from 2% to 26% are requested in the 18 statutory "grades" of federal employees. Grade # 3, for example, would get $85 more yearly and grade # 18 would be boosted $5,270 more (from $21,000 to $26,270).

Even this will not really compete with industry, which, for comparable grade # 18 skills and length of service, pays $38,000. It is therefore quite likely that when the brain pinch really reaches its peak—around 1965—penny-pinching will be abandoned (backed by a new Congressional law) and NASA will openly "buy brains" away from industry.

The author will go out on a limb and predict that the top quality 33% of scientists and engineers—in any category or phase of astronautics—will from 1965 on be such a premium human product that NASA will throw the book away and match industry, dollar for dollar, with the entire U. S. Treasury to draw from. Mindpower, which is still paid *less* than what many blue-collar members of trade unions receive today, will finally be given better rewards, reaching perhaps three or more times the national average for individual earnings.

They won't be overpaid. It will still be only *half* as good as the proportionate pay Soviet scientists receive *today* in their society. It is perhaps one of the oddest commentaries on the social scene that "classless" Russia openly pays its space brains at least five times what the average Russian earns, while "free enterprise" America allows miserly extra rewards for high-powered brainwork which is often worth 100 times its stipend.

Many educators and government leaders feel the time is here when we must shake off our worst form of discrimination—antiintellectualism. As with factory labor a century ago, our mental sweatshops must go so that college professors who perform priceless research, for instance, are not forced to "moonlight" (an estimated 27%) in order to support their families properly.

NASA RECRUITMENT

Starting in 1961, with the expansion of its activities triggered by the Apollo Program, NASA began an organized system of procuring human hardware to produce the space hardware it will need in the decade ahead. Teams of recruiters are sent out from headquarters periodically to scour the country for select brainpower. Some concentrate on big cities or industrial areas and set up a series of interviews with any interested science and engineering personnel in the district. Other groups lay siege to academic centers, inviting undergrads to meetings which reveal NASA's career opportunities, spiced with lectures by many of its most famous space VIP's, such as Wernher von Braun, Hugh L. Dryden and Ernst Stuhlinger.

Each year this recruitment program becomes more intense, and it is certain that any brilliant college student today will be approached by NASA. Applications will also be open to all others, with the understanding that exceptional talent is being sought rather than the average. Jobs at NASA will entail a certain amount of national responsibility and thus the candidates chosen must be of unquestioned high ability. But because top talent is so scarce, NASA has begun to go back to the source and is resorting to the following methods to ease the future pinch:

NASA Training Program

In April of 1962, NASA announced that it "will begin support of a training program at 10 universities . . . in order to increase the supply of scientists and engineers to meet the needs of the national space effort . . . We expect the scope of this program will increase considerably in the years to come . . . *The commodity in most critically short supply is brainpower."*

NASA is going all-out on this important venture. Students chosen for their training program receive generous stipends of $2400 a year plus expense allowances up to $1000 annually. The universities are reimbursed for tuition and other fees involved. Candidates are selected by the universities themselves, based on scholastic record, personal qualifications and the student's proposed research plans. Out of this has branched the Management Intern Program, in which selected liberal-arts students are trained for eventual administrative posts at the various NASA centers.

Again, 322 grads and undergrads were appointed last year as Special Summer Employees at Marshall Space Flight Center in Huntsville, Alabama, to train in booster technology at Wernher von Braun's "Saturn Works." Actual salaries are paid to the students, ranging from $3500 a year to $6345. Most of the selectees are expected to take post-grad studies and obtain at least an M.S., preferably a Ph.D. What with practical summer experience at MSFC each year, the candidates graduate as technical experts ready to step right into an important job in the NASA hierarchy.

Last year Columbia University in New York City inaugurated its Summer Institute in Space Physics for NASA. In it, outstanding students are awarded full-tuition scholarships and are given intensive summer courses in handling the most complex problems of astro-physics. It is understood that those accepting the award will upon graduation join NASA.

These are only a fraction of the various training and scholarship programs that have sprung up, all sponsored by NASA. Throughout the nation, the space agency is planting its academic-trainee seed in order to reap a harvest of golden brainpower in the critical years ahead.

NASA's "headhunters" are foraging most heavily among high-schoolers, lavishly supporting Science Fairs and similar programs. It is here that they have dropped their usual mask of scientific reserve and have presented their true face, seeking to inspire fresh young minds:

"As the frontiers of the Universe are penetrated, mysteries will be unveiled, undreamed-of wonders will be revealed, knowledge and understanding will be increased, and mankind will no doubt reap innumerable benefits as the cosmic curtains are rolled back."

They are speaking to *you*, the raw material out of which America's future scientists and engineers will be molded. They mean every word, more intensely than you can know. Without you, those cosmic curtains can never be unrolled.

Or listen to the words Astronaut John Glenn spoke recently, in his quiet but impelling voice, to a convention of high-school administrators: ". . . curiosity is a wonderful thing. Is there any reason why our learning curve (for young students) must reach a plateau?" Glenn went on to seriously state that it was the task of educators to put the "exploding knowledge of the Space Age into the minds of America's youth."

At the same meeting, Major Robert White, famed Air Force

space pilot who took the X-15 rocketplane 60 miles high at a blistering 4000-plus mph, heartily agreed with his fellow spaceman, and added: "The gateway to space is through the doors of our schools. . . . There is no such thing as too much knowledge, and only Heaven knows what one will need, or find useful, on the hard climb to the stars."

They ought to know. They have started that climb. And it is up to you, the younger generation, to climb higher on the ladder of space, if not in person, then with something quite as vital—the booster thrust of your mind.

In the space program, you can in spirit be a John Glenn or Bob White, pushing back the frontiers of the universe.

7 : NON-INDUSTRY
ASTRONAUTICS AREAS

*(Pure research in universities . . . Non-profit research founda-
tions . . . Armed services colleges . . . Technical training in
Army, Navy and Air Force schools . . . Growing opportunities in USAF astronautics . . . The teaching profession . . .
Faculty salaries . . . Need for science instructors . . . Untapped
mindpower of women . . . Noted women technologists . . .
Anti-female discrimination in technical fields . . . Campaigns to
enlist scientific womanpower.)*

ACADEMIC RESEARCH

In institutions of higher learning (beyond high school) many
teachers have a special "side line" of prime importance to the nation—research. Up to 250,000 college instructors carry on some
form of research, occupying anywhere from 10% to 95% of their
time, depending on how many classes they must handle daily or
weekly. However, less than 100,000 can be considered first-rank
researchers, and this creative core—equal to a football or baseball
audience at one stadium—is the prime source of America's march
into the unknown.

Most of them—in sharp contrast to all other researchers in the
armed forces, NASA or industry—follow only their scientific curiosity, seeking basic fundamentals. Their paths of investigation
lead into the microcosmos (atomic science, molecule structure,

viruses), the macrocosmos (origin of the universe, galaxies, gravity phenomena) and every stop in between.

It is creative work that separates the genius-grade minds from the merely great. Among you school-going citizens of today may exist an Einstein or Madame Curie, whose basic discovery-to-come will have the fantastic power to affect America and all the world. Even you who are semi-geniuses will profoundly change the course of future civilization. It doesn't matter how utterly "non-worldly" or "useless" your fundamental find seems to be.

When Dr. Harold Urey discovered deuterium (double-weight "heavy" hydrogen) some 20 years ago, he himself announced flatly it would "never have any practical applications." When deuterium proved to be the key to thermonuclear power (fusion principle), Dr. Urey wryly observed that his earlier prediction was the "greatest mis-estimate" of all time. Despite the initial use of this awesome energy in the H-bomb, the future promise of the thermonuclear reaction is far more important as an answer to the world's needs for unlimited peacetime power.

Certain universities house world-famed research facilities—Jet Propulsion Laboratory of Cal Tech, Lincoln Laboratory at MIT, Applied Physics Laboratory at Johns Hopkins, and a dozen others. More recent phenomena on the scientific scene are the non-profit research organizations run by themselves rather than by a university. Among these are Battelle Memorial Institute, Rand Corporation and Armour Research Foundation, which carry out excellent "pure science" investigations as well as applied-science studies. Their staffs too are given free play in delving into scientific depths never penetrated before.

Any of you who inwardly thrill at scientific exploring far off the beaten track will find your niche in these university or endowed laboratories. Don't be surprised if your "ivory-tower" discovery is immediately snatched up and turned into practical use before you finish writing your final report. The gap between finding and utilizing new science principles is fast closing, in this hungry age where technology is right at the heels of theory.

If you qualify, you can get into this area simply by staying at your college or university, after graduation, as a teacher. In time, when you have moved up the faculty ladder, you will earn research privileges which can eventually take up most of your time if you so desire.

MILITARY ACADEMICS

Through its three armed services, the Department of Defense also offers technical education and job opportunities to not only civilian employees but to its uniformed personnel in the Army, Navy and Air Force. If you are of draft age and have little chance of deferment, this will be of particular interest to you. The problem of whether to enlist or to wait for the draft can perhaps be best answered by visiting the armed services recruitment offices and talking it over. They will give you sound, objective advice best suited to your individual case.

It is perhaps not too well known that the armed services have since World War II recognized the need for skilled technical personnel and today maintain a vast pool of educational facilities. College degrees can be obtained, provided you sign up for an extended career in uniform at an officer's rating. But many non-degree technical educational plans are open for those who will return to civilian life as soon as they have served their term. Out of some 2,500,000 men in uniform, no less than 640,000 are today receiving schooling in every phase of science technology.

Army. Each year, over 150,000 GI's graduate from 35 Army schools, offering 500 courses, that are scattered all over America and the world. Many of these men are processed through the Department of Ordnance, which trains men in the technicalities of military missiles and space rockets. Many variations are offered to the GI—2-year technical courses similar to civilian vocational schools, correspondence courses, part-time civil schooling at Army expense, co-op education plus job training, and so on.

Any competent GI may apply for one of these opportunities to gain an excellent extension of his education, which can be of great help to him when and if he returns to civilian life. The Army gains, in turn, by constantly replenishing its vast pool of technicians to keep its technological pursuits well staffed.

As one example of Army academics, its Guided Missile School at Redstone Arsenal, Alabama (near the MSFC Saturn works) lists some 60-odd courses in missile systems and trains over 2000 student-GI's each year. The atmosphere at GMS is astronautically inspiring, for it was here that space history was forged from 1958 to 1960 when von Braun's Jupiter-C first put America into space (Explorer I) and hurled a payload past the moon (Pioneer IV). Redstone Arsenal is the Army's center for new missile development (Pershing, for instance) and is known as the "Home of Inventors" for its many brilliant breakthroughs in rocket technology.

Also well-known is the Army Aviation School at Fort Rucker, Alabama, which processes over 6000 students yearly into pilots (Army helicopters and transport planes) and aeronautic technicians. Many of these are then eligible for choice berths in the Army system relating to missilery and rocket technology.

Other Army schools specialize in radar, telemetry, telecommunications and many more Space Age techniques. Though today it has no space project entirely its own, the Army has an official joint role in satellites for space communications, weather observation, geodetic and mapping services and maintenance of a future moon base when that day comes.

Thus, you can turn your hitch in Army uniform into a "prep school" period and lose very little in going on to your college degree after discharge. In many cases, technical Army training sets you up as a skilled technician in some specialized field of rocketry or space technology, so that you can step into a good job the moment you shed your uniform.

Navy. The Navy too (plus Marine Corps) maintains dozens of technical schools and academic installations of excellent calibre. The Naval Research Laboratory in Washington, D. C., is noted for creating the specialized "piggyback" satellites (Greb, Lofti, Injun, etc.) that are orbited along with its famed Transits (ship guidance satellites). Other Navy space-science facilities include the Naval Ordnance Test Station at China Lake, California (guided missiles) and the Point Mugu launchings from PMR.

As in the Army, young sailors with the right aptitude can apply for all degrees of training in electronics, metallurgy, rocket-fuel technology, Polaris guidance systems, and other broadband techniques covering the rocket and space field.

Air Force. The opportunities are almost unlimited for you upon donning the USAF uniform. In fact, the Air Force's large recruitment force of 1300 men constantly combs the nation to entice young enlistees into its technical programs. These efforts, plus an in-service brain hunt, result in 100,000 men and women in AF blue graduating yearly from their many technical schools.

These include their own colleges—a dozen besides their famed Air Force Academy, which offers degrees and officership for life. There are also many non-degree technical schools molding bright young minds for the USAF's enormous space program, second only to NASA's.

And it is conceivable, if a Soviet military space threat materializes, that the future may see the USAF dominant over NASA in the over-all space effort simply for national security—or last-ditch survival.

But even under the present balance of NASA-dominant space leadership, the second-place USAF offers its uniformed personnel a wide choice in Space Age training programs, some far older than any NASA establishment. Job assignments deal with giant ICBM rockets, scientific payloads (as in Discoverer satellites), the X-15 facility, and such combined USAF/NASA projects as Gemini and Apollo.

And in bio-astronautics, the USAF is unchallenged. First organized in 1949, its famed Department of Space Medicine (Brooks AFB, Texas) has made history with its man-versus-space studies: experiments with simulated spacecraft, training "chimponauts," and planning far ahead to the time when our astronauts will be orbiting all over the solar system—perhaps beyond.

Mufti Careers. If you are subject to draft (or enlist in advance), all of the above training is available in the armed forces and should certainly not be bypassed. Instead of merely "marking time" while in uniform, you can actually further your civilian career in the space field, or come out as a ready-to-work technician for whom jobs galore are waiting. You may even decide to remain in uniform and join the armed forces' never-ending space effort for life.

ACADEMIC CAREER FIELD

One important space career area, apart from civil service, the armed forces or private industry, is that of teaching and/or research for colleges and universities.

The teaching profession does not produce astronautics hardware, but it does produce astronautics brainware. Teachers launch engineers and scientists who in turn launch space vehicles. Hence, in its primary role for the long term, teaching is the *most* vital aspect of our space program. Again an entire book could be devoted to this subject alone, and only a brief rundown can be offered here.

Teachers are by far the largest single class of professional people in America today, totaling over 2,200,000 (elementary, secondary and college, on both part-time and full-time basis). Second are engineers, aggregating something over a million.

However, of the total teaching force, only some 300,000 conduct classes in science as their primary service (mostly in colleges). But because of the population explosion—which will severely strike all colleges by 1965 as the wave of "war babies" arrives—the threat of a technical teacher shortage also looms.

Already the colleges and universities themselves are vying to keep their own scientist/engineer graduates as faculty members.

Teacher "feedback," as they call it, has become a serious academic problem. Pay is notoriously low, compared to industry, but has improved markedly since 1960 and is still steadily rising as needs for teachers grow.

Starting salaries for teachers in all public schools (elementary plus secondary, 12 grades through high school) range from lows of $3200 (a few southern and middle plain states) to $6200. Average comes out at $5527. In numbers, 6.4% average below $3500, 28.8% between $4500 and $5500, and 8.6% over $7500. The latter are in the big industrial states, such as California, Illinois, New York and New Jersey.

Of the more than 1,000,000 elementary school teachers (public, private and parochial) 86% are women. Among the 743,000 high-school teachers, 55% are men. Men also comprise 75% of 533,000 teachers in higher education.

Taking colleges only, the nationwide median salary (adjusted average) for full-time teachers (men and women) is $5095 for instructors, $6231 for assistant professors, $7332 for associate professors, and $9107 for full professors. Salaries tend to range up or down considerably depending on the size of the institution and its endowment resources.

Also, the aid program given to state colleges by their local educational boards varies widely, following the better-pay reaction that came along when the low pay of teachers took the limelight several years ago. Some states have pushed through progressive high-pay programs for teachers, while other communities still slumber in pretechnological policies as outdated as the sweatshop.

The Space Age, demanding strong emphasis on curricula in the physical sciences and engineering, has spurred many universities to build up competent technical faculties—and the only way to get the best is to pay the best. Harvard professors, to give one example, average some $17,000.

For the future, it is almost inevitable that technical teachers will be paid much better than by today's standards. The NEA (National Education Association) predicts that by 1970 *average* pay (not median) for college teachers will reach $9710, compared with $5800 in 1963. Local school boards so far have been slow to recognize what a shoddy state our neglected school system has fallen into, in plant as well as in teaching talent. They are not yet bidding at the right prices for the nation's brain pool. But being satisfied with educational mediocrity is impossible for the future. The space pressure is on.

Eventually, when worn-out educational traditions and tax-worshiping prejudices break down, federal aid to all schools will be accepted as not only necessary but proper. The government is prepared at any moment to pump in as many billions as required. Only a 19th-century "state's rights" attitude among self-centered school boards dams (and damns) this flood. As most modern educators know, in or out of the government, education is not an *expense* but an *investment*. New billions poured in (and a federal dollar is as good as a state dollar) can double our mindpower resources and quadruple our technological capacity. From time immemorial it has been proved that the nation which passes on knowledge the best to the next generation prospers the most.

But as with the underproduced professionals who join the outside world, the left-over teachers of science and engineering are scarcer than first-water diamonds. The nation's existing force of technical teachers (with *degrees*, as distinguished from normal-school trainees) includes only 78,000 in elementary grades and high schools, and 130,000 in colleges, for a total of barely over 200,000. By 1970, it is estimated the country will need 648,000 more. But we are falling short by 50,000 a year. Obviously, if this teaching gap is not closed, our entire space program will dry up at the source and getting to the moon will literally be "reaching for the moon"—hopelessly.

Yet everything has two sides, and despite inadequate pay today, the teaching profession does offer several desirable fringe benefits no other field does:

—College instructors average 12 to 15 hours of classroom work a week, leaving time for research, writing or other scholarly pursuits.

—Full professors, with as little as 6 classroom hours per week, often work on spirit-challenging research grants awarded by the space program.

—Summer vacations of three months are automatic, plus sabbaticals and leaves-of-absence upon request. It is optional to teach summer courses at extra pay, or work at "moonlight" jobs outside of academics to augment income.

—In cultural and academic circles, if not everywhere, teachers are accorded honor and prestige for what is certainly one of the noblest of professions.

—Teaching, or passing along the priceless gift of knowledge to the next generation, shares with scientific research the distinction of rewarding its devotees with the greatest acme of self-satisfaction known to the human soul.

For those of you who are technical-minded, and also inclined toward a secluded life that features the special attractions above, teaching pure or applied science in college (or the lower grades) is no less important than making space hardware. Each young mind you inspire to aim for the stars has the potential of being a new astro-science Einstein or space-engineering Columbus.

UNTAPPED MINDPOWER

The teaching field, one of the few professions in which American women have played a large role, brings up a new point. This book is not aimed at "men only" but *minds* only. And when it comes to brainpower instead of musclepower, sex drops out of the running. Obviously, 50% of our nation's mental brilliance is possessed by our feminine citizens—yet *less than* 3% is utilized in the science technology field, though women make up 30% of our total national labor force.

All down the mental assembly line, the female mind is conspicuous by its scarcity. Only ¾ as many girls as boys finish high school. Of the high-schoolettes that do graduate, another ⅓ fails to go to college. And once in college, to compound the cranial crime, American women shun the technical fields like poison— mostly because American men's attitudes discourage them.

Why does this happen? Because it is a tradition in American society that science and engineering are "he-brain" professions. The facts show otherwise, wherever women have had the courage to invade this male-dominated field:

—Ann Eckels Bailie, a mathematician handling computer data for our tracking system, discovered from analysis of Vanguard I's orbital quirks that the earth is pear-shaped.

—Helen Mann, another computer specialist in tracking, works at Cape Canaveral as an Impact Predictor during each launch— determining where a Mercury capsule will land either if it fails to achieve orbit, or after the first and any succeeding circuit of earth.

—Nan Glennon, first girl engineer of the University of Southern California, is an invaluable addition to the management staff of Space Technology Laboratories.

—Edith Olson, chemist, received DOD's highest civilian award for devising a type of printed circuit that shrank miniaturized payloads of satellites even more, down to micro-miniaturized proportions.

—Mary Romig, re-entry mathematician for Convair, deals in studies of Space Age speeds beyond Mach 10 (6600 mph), which will apply to the Dyna-Soar's landing maneuvers after a space jaunt.

Women technologists like these carry scientific workloads as heavy as any man can tote. NASA has wiped out all anti-female prejudice and eagerly hires the distaff doctors of science:

—Dr. Nancy Roman, Chief of Astronomical Applications, in charge of the OAO satellite project.

—Eleanor Pressly, expert on sounding rockets at Goddard Space Flight Center.

—Harriet Malitson, solar physicist at GSFC.

—Marjorie Townsend, top-grade electronics engineer in space systems.

—Ann Bailey and Marcia Neugebauer, full-fledged scientists of NASA's in-house facilities.

NASA has over 250 women classified as professional aerospace science technologists and mathematicians. Bell Telephone Laboratories, perhaps the world's greatest source of major breakthroughs (transistor, solar-cell, radio-telescope, MASER, traveling-wave tube, Echo ComSat and Telstar microwave relay systems) has 400 scientists who wear skirts.

At the First International Women's Space Symposium (Los Angeles, February, 1962) guest speaker Captain Robert Truax, famed rocket-propulsion expert of Aerojet-General, rejected the "Kinder, Kürche und Küchen" philosophy regarding women and coined the phrase: "A Woman's Place is in Space." He was dead-serious, and pointed out that only womanpower added to our manpower can enable America to win the Space Race. The stark fact is that 29% of the Soviet engineer graduates are women —*37,500 a year.* America produces 3,000.

The leaders of the Women's Space Symposium pledged themselves to arouse America against this senseless waste of brainpower, and to spearhead a nationwide movement that will break down the brainwashing traditions which discourage brilliant females from entering the one profession that can truly use their latent mental powers.

There are several entrenched beliefs held by male leaders in business, government and academic institutions that must first be smashed:

—That women "don't belong" in the "man's game" of free enterprise. (Dr. Beatrice Hicks runs an electronics firm and runs her male-led competition ragged.)

—That women are too "flighty-brained" to concentrate on

"deep" science or engineering problems. (Laurel Roennau, bio-astronautics engineer, has evolved life-support concepts that trousered space-medicine experts use as unexcelled reference works.)

—That with marriage and raising families, women can only devote themselves to a "short career." (Past statistics on working women in general prove that the 25 million female job-holders today will put an *average of 25 years* into their earning careers.)

And so on down the line. A common attitude is that women are too "delicate" by nature to handle tough jobs without collapsing emotionally. In Russia, 70% of the doctors—including surgeons performing major operations—are women. And were our pioneer women of the last century, who fought Indians and felled forests alongside their men, of a more *superior* breed than their descendants today? To believe this is to deny the total process of evolution, or to claim that world records falling to modern women athletes are falsified.

And if American he-men cling to the delusion that she-women can't take it—mentally, physically or intestinally—let them look at the records of thirteen "Astranette" candidates who qualified for future spaceflights. In the brutal tests—high-g centrifuge whirls, simulated space isolation, "idiot box" sessions with flashing lights—they scored as high as the men, except where they scored *higher* (in endurance of heat and cold, keeping alert after long routine tasks, and reaction time to danger signals).

It is obvious that feminine "inferiority," with the sole exception of muscular HP, exists only in male (and perhaps misguided female) minds. One proof of the technological potential of woman-power is the fact that girl college students are winning over 40% of all jobs for technicians. Scientists and engineers with girl aides declare them "ideal" assistants, often superior to male equivalents. The reason is not hard to find. They come from the ranks of the women with undeveloped scientific aptitude who each year fail to obtain a degree. First-class female minds therefore become mere technical "assistants" to men in laboratories, when in many cases it could probably be the other way around. One can only estimate how many "Madame Curies"—lost because fathers or male advisors frowned them out of full professional training—are today wasting their true talents in lesser jobs.

Still, becoming technicians currently represents the golden way for gifted girls to at least partly do their scientific bit, until the day comes when America as a whole sheds its crippling anti-woman attitude in the professional fields.

One other and very major barrier to recruiting female mindpower is wage discrimination. In almost all fields, including science and engineering, unfortunately, women are automatically paid less than men under comparable circumstances. They are also held back from rightful promotions into jobs of responsibility and challenge, purely under the "unwritten labor law" that women can't do a "man's job." Countless documented exceptions prove this to be a female fairy tale, on a par with the legend that men can't cook as well as women.

Though comprising a majority (52%) of the population, women are actually the most exploited "minority" in America today. They gained the vote but are still "second-class citizens" in the pay scale of the nation. And no area of racial or economic discrimination is as irrational as the anti-feminism that grades brainpower according to sex instead of performance. The Fair Sex is being badly mistreated by the *Unfair* Sex.

America cannot afford this bigoted "Battle of the Sexes." In the Battle of Brains with Soviet Russia, keeping our women doing second-rate work will ultimately guarantee a third-rate America in space and a fourth-rate one on earth.

That is why everything in this book is meant for you, Miss Student USA, as well as your brothers. Each year, 50,000 of you are "mental drop-outs" who even in college train your high-IQ brain to fulfill a low-IQ life devoid of satisfying scientific pursuits. But fortunately, even if slowly, the "girl, stay home" prejudice among unmasterful males is breaking down today, under the impact of the technological age and its brainpower demands.

The first girl engineer graduated last year from the University of Michigan—with the highest honors of her all-male class. In college after college, enrollments of women in technical curricula are gradually increasing. NASA and a growing number of space firms are on your side, giving equal opportunity for brainwork without a gender label. You won't find their pay envelopes marked "his" and "hers" and containing different payloads.

Even if the rest of America takes longer to catch up, the astronautics want-ads are rapidly putting out the welcome mat for women with the motto: "There is a space in the Space Age for you."

Like young men in school today, you co-eds can have a space career for a lifetime, or any part thereof. As missile maids, you can make joint history as the men and women behind our spacemen. And eventually, behind our space*women.*

8 : THEY WANT YOU

(Recruitment of college graduates by firms . . . Salaries and inducements offered . . . Better grades, better jobs . . . On-the-job training or "Learn and Earn" programs . . . Engineer, scientist and technician totals in industry . . . Geographical distribution of aerospace firms . . . Projected growth of key space-industry complexes by 1970 . . . Definitions: prime contractors, sub-contractors, supply firms . . . Contrast between R&D and production . . . Feasibility studies . . . Major firms in the aerospace field . . . Increasing number of firms climbing on the space bandwagon . . . Is astronautics already America's #1 industry?)

You are in your senior year at college. You will get either a science or engineering degree on graduation day a few months ahead. Your marks are good, within the top 10% of the class. You wonder just where and how you will get a job when you step out with your sheepskin.

You don't have to wonder long. Recruitment teams from aerospace firms are already camping at your doorstep.

By arrangement with your dean or the faculty, these representatives of various giants in the astronautics industry are allowed to contact you and interview you and other promising students in your final year. Some of the bigger space firms have more than 10,000 scientists and engineers on their payroll and desperately need more. They therefore shop among each new crop of fresh college grads. This brain hunt is carried on at all the larger colleges and universities by up to 100 big space firms.

Each company recruiter strives to capture the prize pupils with the best scholastic records, personal qualifications and creative abilities. If you are among this educated elite, you will undoubtedly receive offers from several firms and thus have a choice of jobs.

In vying for the exceptional college prospect, each company recruiter will usually make special concessions to win you away from his rivals. It may not necessarily take the form of higher salary, but can be other inducements—finding a place for you to live before you arrive at their plant, an extra week's vacation the first year, paying all transportation costs if you happen to be married, giving you a temporary loan or whatever else will benefit you personally.

However, the level-headed grad will not go so much by these relatively unimportant *short-term* gains, but will look to the *long term*. Which firm best suits you? Factors that might sway you one way or another are how heavily a firm goes into research, what kind of space vehicles it makes and whether its general labor policy is favorable to your future or not.

Recruitment usually starts with correspondence from a firm, requesting an interview with you. During the first brief approach, the recruiter will give you general information about his firm, the kind of work you can expect to do there, the salary range and the like. You, of course, are allowed to ask any questions to get a clear picture, but it is likely that there will be more interviews before anything definite works out. If the firm is strongly interested in you, they will want a final commitment before graduation day. You cannot expect to keep them "dangling," with the chance of losing you.

You must bear in mind that though you are desirable "goods" you cannot "play" the firms against one another without losing their patience and respect. It is all right to ponder over several bids for your services, but not to look for the "jackpot." Your college counselor will give you advice when you are in doubt about which of several offers to accept.

This refers mainly to four-year B.A. or B.S. grads. Recruitment of M.S. and Ph.D. candidates is somewhat different. Here, the firm's representative can offer more liberal inducements because of the advanced degree. In the case of a Ph.D., the candidate can actually name his salary range within reason, and also stipulate the conditions under which he will do research.

The tiny pool of yearly Ph.D. graduates in science research and technology (under 10,000) means fierce competition for their services. As yet, firms obtain only a minority of them, because there are more opportunities for pure research in universities or government agencies, where aims are less "commercial."

In effect, industry says to the Ph.D. researcher: "Pick any field you want as long as it fits ours." This is not precisely applied research, which names the specific goal (a better nose-cone for re-entry, for example), but it is uncomfortably close to it. Yet because of the growing sweep of astronautics, in which even applied research means crossing into the frontiers of pure research, the proportion of Ph.D.'s joining industry is growing every year.

However, industry recruitment is most active among the M.S. and B.S. grads. The top 10% scholastically, at any accredited college, are prime targets and seldom reach graduation day without being signed up or optioned for a job at maximum starting salaries. The next group—within the top 33%—are still sought after and in many cases obtain a good offer from some aerospace firm. From there on down, the pickings are leaner for both firm and grad, though some members of the lower 50% are hired by firms who are rapidly expanding their technical staff at the time and need all grades of engineers and scientists, brilliant or not.

Not that the rest—those not snapped up by the recruitment system—will be jobless after graduation. In fact, remaining jobless for long would be quite a job for you in the face of the chronic mindpower shortage in the aerospace field. A glance at want-ads in technical trade journals, or even in newspapers, and you will see dozens of firms pleading for *any* young man of scientific or engineering standing to apply.

Somewhere along the line there will be a job for you. Few if any of the pitiful handful (55,000) of technical experts our colleges turn out each year will be without a job shortly after graduation, if they look around at all.

Also, various employment agencies have waiting lists of jobs for the tech grad. In fact, engineering societies, government agencies and non-profit organizations have set up many special services to get the trained brain and the talent-starved firm together. Your college counselor can give you a list of these agencies, which have offices in most big cities and regional areas.

Even if you are in the *lowest* 10% of your graduating class, but have that document naming your degree, you can still qualify for a good job as a technician at the higher echelons. In some cases, their pay is equal to that of starting engineers (about $575 a month).

JOB CHANGING

What if, after you start work for some firm, you find that for personal reasons you are in the "wrong" job? Don't let it worry you.

It happens to perhaps the majority of young engineers or scientists, and there is a rather large turnover of new help among the firms. Statistics indicate that the average young engineer changes jobs and firms 3 or 4 times before he finds his true "haven" and settles down. It is impossible in advance for the firm's recruiter to tell you exactly what kind of work you will do, whether you will be under a good boss or not, what the living facilities are nearby, etc. All these variables may add up wrong for you, after you start.

The answer is simple—seek another job.

But before doing this, be very honest with yourself. If you are

looking for something "easy," or high pay with little responsibility, or rapid advancement before you even prove yourself,
then changing jobs won't help. In that case it's you, not the job,
that's gone "sour." The plain truth is that science and engineering
are among the *toughest* careers going. On special rush projects
that are important to our national space goals, you may be asked
to work long hours and flog your tired brain to do the "impossible" until the crisis is over.

Get it out of your mind that any job in the astronautics field
is "routine" or "leisurely" or a "cinch." Our whole space effort
has been on a "catch-up-to-Russia" basis from the start and that
pressure won't ease up for years.

You may come to envy your classmates who chose "sensible"
careers in other fields and work like men, not dogs. But they
will eventually envy you for the spectacular projects you will be
identified with, and they'll never know the enormous pride you'll
have when "your" rocket or space vehicle rams its way into space.

If you're afraid of challenges that each time will demand twice
what you gave before—or twice what you thought you had in you
—then you shouldn't even try to be an astronautics pro. Space
separates the "supermen" from the men. And if that term seems
out of place, you have only to read the details of a countdown
crew of experts nursing a giant rocket to its lift-off, often laboring day and night with little or no sleep and endlessly fixing
hardware flaws, to know that they *are* supermen. If anything can
call out the best in you, it will be the relentless goading of space
work, which is 99% pioneering in the new frontier's wilderness
and 1% routine.

From the impossible, you go to the incredible, after which
comes the hard stuff—the inconceivable.

Make no mistake. All the miracles accomplished in the past five
years of the Space Age are child's play to what the next five years
will demand from its toiling band of space invaders. Sweat and
blood will really fuel the rockets you work on, but you'll earn

your place in history as one of the superminds behind the space-men.

Future feats aside, just what do you do at the *start* of your space career here in the present? You're just an "engineer," who is use-less as such with only his schoolbook experience. You need to know some one of six dozen specialties in space work, but how do you gain experience?

Most firms have this problem solved for you, from their own long experience with professional help. You are first given a choice of broad fields, to find the one that most suits your tem-perament and abilities—R&D, designing, production, testing, evaluation and the other ways in which applied scientists and engineers can serve the over-all technical system. In some firms, job-aptitude tests are given and the conclusions are compared with your own leanings to make sure you won't be a square payload in a round rocket.

Your choice made, you are then assigned as an assistant to an experienced engineer doing that kind of work, or you join a team of pros involved in that type of project. For weeks or months, however long it takes in each particular case, you just observe and learn. Gradually you are given actual work to do in small doses, until you are ready to be on your own.

Some firms send their brand-new engineers to a technical school (or have their own classrooms) for a short, intensive course in some technological specialty. Fees are usually paid in full by the firm, plus your regular salary.

This learn-and-earn system is carried even further if a B.S. man shows the desire and aptitude for more academic training. He may then be enrolled in a college full-time for one or two years to gain his M.S., or the arrangement may be part-time schooling as well as a certain number of work-hours per week for the firm.

In any case, the company is willing to invest this extra training cost in you if you respond and show promise of becoming a valuable future addition to their top-flight technical roster.

In essence, the space firms want each brain they hire to reach its *full* capacity. It is the only way to stretch out the critical shortage of brainpower facing the industry as a whole. And if you are willing to cooperate with them, they are not only willing but eager to reciprocate and see that you use every individual talent you have —plus others you never suspected you had.

The astronautics industry, as mentioned before, snatches up the cream of science and engineering talent as fast as it emerges from college. As of today, out of the all-industry total of some 1,300,000 technical professionals, about 175,000 engineers, 45,000 scientists and 157,000 technicians work for aerospace firms. These include all R&D outfits, airframe makers, rocket-fuel producers, electronics specialists and manufacturing plants that deal mainly or significantly with missiles, rockets and space vehicles.

Their contracts include awards from DOD for military missiles as well as for space projects of the USAF and NASA—namely, both the *defense* and *space* hardware that makes up the sum total of rocketry. Of the entire space melon of some $12.5 billion in fiscal 1964, over $10 billion is allocated to private industry, including about 100 prime contractors, 1000 sub-primes and 200,000 suppliers. Thus the space money that flows from the government's military and civilian agencies eventually spreads throughout these 50 United States to every concern, from giant corporations down to small independent companies employing a dozen people.

Obviously, the highly industrialized areas—New England, East Coast, Midwest complex, West Coast—get more than the rural regions of the Plains States, Mountain States and South. The South, however, is gaining rapidly, as there spreads across it the "space crescent" now being organized for our 10-year program ahead, including the great moon project. Alabama has the Saturn works at Huntsville; Mississippi, the Saturn test facility; Louisiana, the Michoud booster-assembly plant; Texas, the spaceflight center at Houston; and Florida, of course, the famed Cape Canaveral launching site.

Thus, in a huge bow across the southern states, space funds are pouring in and will increase to near-flood proportions during the next decade. Branch offices, labs and plants of the giant aerospace firms also tend to spring up and cluster around such space centers, bringing in more new industry and its golden prosperity for the areas.

Yet it is a single state—California—that has gained the biggest space plums, because its many aircraft makers and early missile plants (Convair, Aerojet-General, Douglas, Ryan, etc) were naturals for converting to the assembly of spacecraft and giant rockets. California takes in a whopping 33% of all space funding. This undoubtedly contributed to its high influx of workers and let it pass New York in early 1963 as the most populated state.

It seems likely that California, with its space hardware complex; Texas, with its future $30-40 billion Moon Project; and Florida, with its "Spaceport" facilities, will be the Big Three states of the coming Space Age.

Estimates that seem optimistic, yet may fall short of reality, predict that 1970 will see: (1) 500,000 new astronautics jobs in California alone; (2) a staff outnumbering the government in Washington to man the Space Brain Trust of Texas; and (3) a swarm of space personnel drawn to a swollen Cape Canaveral that will make Florida the third most populous state (after California and New York) in the Union. Lesser but still important space industry segments of America will be "Electronics Row" near Boston, the huge New York-New Jersey "industrialopolis," the Detroit-Chicago mill and plant complex, the nuclear and missile R&D/test ranges of the Nevada area, and other facilities throughout the nation.

Yet in time, the country as a whole will partake of this vast new super-industry in varying degrees—and already is. As of today, at the mere beginning of the Space Age, many firms from large to small have joined the space bandwagon. They are divided into groups as follows:

Prime Contractors. Often called just "primes" in the trade, these are the big firms who are able to take responsibility for the production of a total rocket or vehicle. Though some of the hardware is actually made by them, most of it is sub-contracted for elsewhere. The prime's main job is management of the full project, assembly of the many parts, testing for reliability, and final delivery of the operational product. Convair's Atlas booster, Lockheed's Agena upper-stage vehicle and RCA's Relay satellite are typical examples of primes and their comprehensive projects.

Sub-Contractors. These are "sub-primes" who take over large portions of the R&D or hardware for a specific project of a prime. For example, Rocketdyne produces and delivers to Convair the complete engine system for the latter's Atlas missiles. Also, Aerojet-General provides the motors for the Titan to its prime, Martin-Marietta.

Other firms which take on sub-prime contracts are General Electric (satellite instrumentation), Sperry-Rand (guidance systems), IBM (trajectory computers), Avco (nose cones), Thiokol (X-15 engine) and Pratt & Whitney (Centaur's hydrogen-fueled engine), to list but a few of the 1000 in this category.

Suppliers. Some 200,000 firms in turn receive contracts from the sub-primes for the 200,000 to 1,000,000 parts that make up a complete satellite, probe, upper-stage carrier vehicle or booster rocket. An average 40% of gross funds for any space project goes to small business firms (under 500 employees) scattered throughout the 50 states. They most often supply just one or a few specialized gadgets or items—printed circuits for satellite telemetry units, beryllium bolts holding rocket stages together, accelerometers, horizon-scanners, radar components, space-suit fabric, squeeze-tube astronaut foods, transistors, solar cells, oxygen pumps, electric relays, timers, ignition squibs and a myriad of micro-miniaturized parts and devices.

Supply firms run the gamut of industry—steel-makers, electronics labs, chemical plants, wire-manufacturers, glass-makers, plastics-producers. You name it, space uses it.

IDEAWARE TO HARDWARE

The money allocated to aerospace firms by NASA and the USAF is split into about 30% for R&D, 70% for workable hard-ware.

Even though rockets, upper-stage vehicles and spacecraft are be-coming operational in increasing numbers, R&D continues to take its huge bite, simply because new hardware designs are constantly on the drawing board for the future. It takes anywhere from two to ten years to develop any particular large item—missile, rocket, booster, upper-stage, satellite, interplanetary probe, manned ve-hicle—from concept to countdown.

The Atlas first became a firm project in 1952 and produced flight-test hardware by 1957, a span of 5 years' development. The Mercury capsule of the astronauts, less complex, took only 3 years, from 1958 to Shepard's first manned flight in 1961. The Rover nuclear rocket, however, may stretch from its first Kiwi engine prototypes of 1959 to 1967-69 before an operational model is ready to go.

However, new concepts for any type of space hardware do not spring instantly into R&D. Preceding this is the *feasibility study*.

Teams of brilliant scientists and engineers in every firm, and in NASA/USAF laboratories, first dream up new *ideaware*, without form or substance except as a few notes and hasty sketches. If the new concept seems worth pursuing, it is submitted to a detailed evaluation by science engineering experts, out of which blooms *paperware* that includes diagrams, charts, tables and as many pro-jected data as are available.

Often these feasibility studies take thousands of brain-hours and result in heavy batches of paperware—yet are rejected even-tually because some other study proves more promising or there is a shift in the hardware requirements of the space program. Only part of these expensive studies are paid for out of NASA and USAF research funds. Most big aerospace firms carry on their own "advanced space concepts" and foot the bill themselves. There is always the chance of stumbling on something big that the space leadership will want—and the firm can thus win a fat contract for the follow-up R&D contract.

The investigations of advanced space systems go into concepts that are truly fantastic, at least to us today, and will be dealt with in a later chapter.

But leaving the far-out ideas and sticking with the close-in realities of today, the strongest clue to where to seek your space future lies in examining the major NASA contracts awarded to the various firms. No guarantee, naturally, can be given as to whether any certain company, or any currently lucrative field, will stay in the space forefront. There will be unforeseeable downs as well as a general up.

Perhaps the surest bet is that the majority of prime contractors and their sub-contractors, who design and make rocketry paraphernalia, will continue to prosper on the ever-fattening funds fed into our national astronautics effort. They are the "heavyweights" on whom our government depends for all missile and space projects. Their rising gross income in recent years indicates their importance in the drive toward our celestial objectives.

First, let us make it clear that the big "space firms" are not uniform and represent different aspects of the total "aerospace" field, both military and civilian.

Among those which are airframe makers, some still make much of their yearly gross from non-space military aircraft. Typical are the McDonnell Aircraft Corporation, whose bulk receipts come from producing the Phantom and Voodoo jet fighters for the USAF, rather than from Mercury Capsules; and secondly, Boeing Aircraft Company, whose famed jetliners account for greater sales than the Dyna-Soar or Saturn second-stage contracts it won.

Conversely, some former aircraft manufacturers have swung

mainly or completely the other way, notably Martin-Marietta, which made its last plane in 1960 and switched entirely to missiles and space technology after gaining the Titan contract. North American Aviation and General Dynamics are others which are involved in more aero*space* than *aero*space business.

Another class of airframe firms deals mainly in *military* rockets and missile systems for the USAF and relatively little in NASA's spaceware; Douglas is one and Lockheed is another. In fact, the majority of firms dealing in the total field of rocketry have much larger contracts for military missiles than for purely scientific space vehicles; i.e., more money pours in from DOD's $55+ billion defense budget than from NASA's comparatively modest $5½+ billion for space.

Non-Airframe Firms. These include several broad types. Some are "true" astronautics companies whose main business in the past two decades has been earthly missiles and space rockets— Aerojet-General Corporation and Atlantic Research Company are representative.

Then there are the utilities giants (General Electric, Westinghouse, AT&T) and the auto-makers turned rocket-makers (Ford, General Motors, Chrysler), who are reaping bigger shares of the space pie year by year but whose main product, or diversified non-rocket doings, is still their major business. Firms dealing in chemicals, oil, steel, aluminum, construction, plastics and other such products also count only a small percentage of their gross in Space Age products.

SPACE BANDWAGON

However, all these firms are now keenly competing for more of the lustily growing astronautics business, sensing that it may in time top their other sales. Projections of the space budget ahead (NASA only) are now estimated at $8-10 billion annually by 1965 and $12-15 billion in 1970.

Almost all experts foresee DOD/USAF space funding inevitably equaling and then surpassing that of NASA, so that adding on costs for military missiles plus space defense systems may by 1970 push the total missile/space budget to over $35 billion *per year*.

If this seems fantastic, it is on record that as recently as 1959 the leading experts predicted a "staggering" space budget of $3 billion annually—*by 1970*. Nobody suspected that NASA would already pass this mark in 1962-63.

At any rate, big business is going in big for astronautics. Some firms—like Aerojet-General and Reaction Motors (a division of Thiokol)—have been in the field since the start of military missile-making in pre-space times—around 1940. Others have jumped on the spacewagon since 1957. Over 25% of the first 100 aerospace firms of 1963 (by dollar volume) did not appear on the 1962 list at all.

Some industry analysts say the over-all astronautics industry will pass the #1 automotive giant in the "near future." Others calculate that if statistics (which are as yet lacking) included every last factor contributing to missile and rocket apparatus, the space industry would be seen to have already taken over the number one spot in America. The total of astronautics workers has already passed Detroit's 800,000 employees, but this is deceiving in that the highly automated auto industry can produce many more finished cars with fewer personnel than the non-mass-production missile-makers.

But if it hasn't already, the rocketry field is as certain to rocket past the auto-making business as Midas III is to circle the earth tomorrow (the satellite's lifetime: 5000 years). Uniquely, for the first time in history, a new industry's shining future can be predicted with utter certainty. The Space Explosion can no more be halted than a nuclear bomb after the chain reaction has started.

Join any of the aerospace firms if you want to enjoy a Career Explosion.

9 : SPACE BIG LEAGUES

(*Corporations dominant in the space field* . . . *Big Three, Top Ten* . . . *A hundred space specialists* . . . *Moon bandwagon* . . . *Apollo team of industries* . . . *Space projects under long-term federal funding* . . . *Big four primes in lunar project* . . . *Apollo sub-primes* . . . *Your job guide among space giants* . . . *Profiles of North American Aviation, Inc., Douglas Aircraft Co., The Boeing Co., Chrysler Corporation.*)

For several years in any list of the biggest Space Age contractors, there has been a Big Three, who sometimes change places but always stay on top.

This aerospace rating is by total DOD/NASA government contracts only (including military jet planes), for fiscal 1962 (July 1, 1961, to June 30, 1962):

1. *Lockheed Aircraft Corporation.* DOD: $1,419,500,000. NASA: $4,951,000. Total: $1,424,451,000.

2. *North American Aviation Inc.* DOD: $1,032,500,000. NASA: $199,109,000. Total: $1,231,609,000.

3. *General Dynamics Corporation.* DOD: $1,196,100,000. NASA: $27,937,000. Total: $1,224,037,000. With jet planes excluded, from 60% to 75% of these aerospace totals goes into missiles and space vehicles.

In actual gross sales, however, including, non-governmental, non-astronautics income, each firm has a higher annual total. GD went to the top with $1,898,500,000 for calendar year 1962. Lockheed reached $1,753,074,000. NAA took third place with $1,633,675,000.

After them, to make up a Top Twelve missiles/space list which has varied from year to year, come more than nine firms. Those that since 1957 have been consistently among the first 12 are Boeing, Martin-Marietta, General Electric and United Aircraft. Others that have been among the Big Twelve at times, but not each year, include AT&T, Sperry Rand, RCA, Hughes and Douglas.

However, if we take NASA contracts only—excluding all military missile work for DOD—we get a different kind of Space Twelve for total contracts in FY-1962. In descending order they are North American, McDonnell, Douglas, Aerojet-General, United Aircraft, Chrysler, General Dynamics, Ling-Temco-Vought, Grumman, General Electric.

Yet other firms have backlogs from previous years on long-range projects and vehicles that would rank them higher than this selection. The full roster of corporations which play prime roles as contractors in our astronautics program today, and will in the future, includes about 100 firms altogether. A few of them are: Space Technology Labs, Avco, Thiokol, Goodrich, Hercules Powder, Garrett, Marquardt, Raytheon, Bell Aerosystems, Goodyear, Schjeldahl, IBM, Republic, Ryan Aeronautical.

Some forty of the top 100 specialize in turning out miniaturized components, electronics gadgetry, telemetry parts, radio beacons, guidance devices, tracking equipment and many smaller items, such as switches, relays, gauges and timers. The better-known corporations in this category are: Western Electric, Minneapolis-Honeywell, General Precision, Texas Instruments, Kollsman, Electro-Optical Systems, Bendix, Northrop, American Bosch-Arma, Electronics & Missiles Facilities, Philco, IT&T.

Another group of about two dozen firms produces GSE—ground support equipment—or the engineering plans that go with it. These firms deal mainly in heavy structural items—radar antennas, launch pads, gantries and service towers, hangars and blockhouse equipment. Some are: Brown Engineering, Rust Engineering, General Mills, Hayes International, Blount Bros. Construction, Litton Industries.

There are several specialty companies who supply services more than hardware. Computer Application Inc. makes over-all cost accountings for specific space projects. Management Services Inc., at Oak Ridge, organizes steps in the Rover Project for nuclear rockets, which the AEC and NASA jointly conduct. Arinc, Arthur D. Little, Santa Barbara Center and others are "business research" outfits that gear NASA's specifications to the contractor's facilities.

One firm is in a class by itself: Thompson-Ramo-Wooldridge, sometimes called a "prime prime," meaning that its seasoned teams of space scientists and engineers first set up a working program on paper before the prime contractor starts the hardware. Most of the pre-digesting not done by T-R-W is handled by non-industry organizations, such as MIT, JPL and various non-profit agencies, particularly Aerospace Corporation.

The above rundown of the aerospace/astronautics industry is unavoidably oversimplified. The wheels of ordinary terrestrial projects multiply into kilowheels within megawheels when it comes to the extraterrestrial craft free-wheeling into the universe. Whole new accounting, management, transportation, inspection and logistics systems have had to be devised for the mechanical beasts that blast away from earth. Even the simplest space booster, satellite or manned vehicle overlaps into ten times the industrial areas that an automobile or airplane does.

THE MOON BANDWAGON

Sometime before 1970, a mighty Saturn C-5 rocket will thunder from its launching pad, propelling an 80,000-pound Apollo-C craft and its 3-man crew away from earth at a speed of 25,000 mph. Destination—the moon. Some 60 hours later, the vehicle will use retrorockets to enter a lunar orbit. One man will stay in the "mother ship" while two astronauts ride a detachable "lunar bug" down to the cratered world below and become the first Americans to land on the moon.

This is the *Apollo Program.* To accomplish this great goal will take a major portion of NASA's space budget in the years ahead, perhaps $30 billion or more. As this massive project got under way in 1962, the largest contracts ever known were let out for hardware. Juicy astronautics plums are falling to thousands of subcontracting and supply firms, but for the prime contractors it is almost "automatic" prosperity for the rest of this decade.

The closest thing to a sure bet, in this modern industrial world, is that the following aerospace firms will expand their plants and payrolls in order to make the moonship hardware NASA needs:

APOLLO PRIMES

Firm	Hardware		Probable Gross Total
North American	Apollo craft		$5,000,000,000
(and Rocketdyne Div.)	H-1 engine	(for Saturn C-1)	
	F-1 engine	(for Saturn C-5)	
	J-2 engine	(for Saturn C-5)	
Douglas	S-IV stage	(for Saturn C-1)	2,500,000,000
	S-IVB stage	(for Saturn C-5)	
Boeing	S-1B booster	(for Saturn C-5)	1,500,000,000
Chrysler	S-1 booster	(for Saturn C-1)	1,000,000,000

Apollo sub-prime contractors with gross totals of between $100 million and $1 billion include Aerojet-General, AiResearch, Avco, Collins Radio, Lockheed, Marquardt, Minneapolis-Honeywell, Pratt & Whitney, Northrop (Ventura), Thiokol, GM (AC Spark Plug), Raytheon, Kollsman, and Sperry.

Thanks to the moon race, those companies are assured of a 5-year-plus inflow of space dollars until the manned lunar landing takes place—and probably after that, as the interplanetary routes are opened up. If you want to take part in mankind's most thrilling venture, join any firm in the Apollo team. As one of the men behind our USA Moon Columbuses, you can make space history.

APOLLO BIG FOUR

Which of the above-mentioned firms should you contemplate working for? Individual factors will of course influence your choice the most, but we can give you aid in the form of a comprehensive résumé of each company.

NORTH AMERICAN AVIATION INC.

3700 East Imperial Highway, El Segundo, California

Profile

Founded 1928. Primarily an aircraft manufacturer until 1952, since when ⅔ of gross income has come from missiles and rockets. Became a billion-dollar corporation in 1957, is now exceeding $1.5 billion, to rank 30th among all U.S. corporations. Total plant space: 17 million square feet. Laboratories: over 200. Total personnel: 110,000, of whom 18,000 are scientists and engineers (including 350 Ph.D.'s) and 10,000 are technicians.

Astronautics Feats

—A Rocketdyne (NAA Division) engine powered the Jupiter-C booster that hurled America's first satellite—Explorer I—into orbit, and sent the USA's first escape-velocity probe—Pioneer IV—past the moon and around the sun.

—Powered by NAA's DM-21 engines, various Thor launch vehicles—Thor-Able, Thor-Delta, Thor-Agena, Thor-Ablestar—have hung over 70 satellites in orbit.

—Astronaut Alan Shepard, first American to sail through space, rode an "Old Reliable" Redstone with an equally reliable NAA engine (H-1).

—A trio of X-15 rocketplanes, history's first piloted spacecraft, was designed, built and flight-tested by NAA.

Company Characteristics

Industry analysts state that a large share of space contracts goes to NAA annually because the firm is *engineer-minded* and pursues *frontier technology*, under a self-financed policy of seeking "tomorrow's technique." Liberal benefits to technical help include in-shop training, rapid job advancement and higher degrees obtainable at company expense.

Divisions and Subsidiaries

Los Angeles Division (corporate headquarters), Anaheim and El Segundo suburbs. Produces USAF jets, X-15 rocketplane and RB-70 prototype (Mach 3 recon/bomber). Employees: 22,000.

Autonetics Division, Downey, California. Aircraft and missile electronic components. Employees: 35,000.

Rocketdyne Division, Canoga Park, California (branches in Neosho, Missouri, and McGregor, Texas). Produces over 75% of all USA space propulsion—for Thor, Atlas and Saturn. Also solid-fuel military missiles. Research contracts in advanced nuclear/electric drives. Employees: 19,500.

Space and Information Systems Division, Downey, California (plus branch at Eglin AFB, Florida). Newest and fastest-growing division, with NASA contracts for Apollo spacecraft and Saturn stages. Also produces USAF Hound Dog missile and Nike-Zeus AICBM cybernetic (computer) system. Employees: 20,500.

Atomics International Division, Canoga Park, California. Nuclear R&D for AEC and NASA, including light-weight reactors for SNAP and SPUR satellite power programs. Employees: 5000.

Columbus Division, Columbus, Ohio. Production of Navy planes (Vigilante, Fury, etc.) plus USAF and Army missiles. Employees: 12,500.

Smaller Facilities

Propulsion Field Laboratory (Rocketdyne) for static-testing giant rocket engines; *Edwards AFB Rocket Test Center* for flight tests of rocketsondes (radio-equipped upper-atmosphere missiles); *Aero-Space Laboratories*, SIS Applied Research center; *Science Center* at Canoga Park (established in September, 1962) for basic research; two *Space Support Engineering Groups* at Marshall Space Flight Center (Apollo Project liaison) and Cape Canaveral (supervising Saturn launch pad under construction).

Main Contracts

Current. All engines for USAF's Thor and Atlas missiles; All engines for NASA space launchers based on above (Thor-Delta, Thor-Agena, Thor-Ablestar, Atlas-Agena, Atlas-Centaur, etc.); Saturn C-1's booster engines (clustered H-1's); Saturn C-5's booster drive (clustered F-1's); J-2 hydrogen-burning engine for Saturn upper-stages; First series of 14 Apollo-A spacecraft, NASA Moon Program; Gemini propulsion unit for orbit-rendezvous experiments; Sub-contracts in military missiles (including Hound Dog, SLAM, Minuteman, Titan, Redhead-Roadrunner, Q-12 Target Drone).

Future (Probable). F-1 and J-2 engine for Nova applications; Apollo-B circumlunar manned spacecraft series; Apollo-C moon-orbit craft; Reactors for Project Rover nuclear rocket; Gemini spacecraft propulsion for USAF to develop advanced man-in-space capabilities; Follow-on versions of X-15 rocketplane for training NASA astronauts and USAF spaceplane pilots; Operational SNAP and SPUR reactors for satellites and spacecraft; Nuclear engines, ion drives and electric propulsion for megapound payloads in orbit, and for multi-ton spaceships throughout solar system.

Technical Help Wanted

Like kindred firms, the NAA astronautics empire is on a perpetual brain hunt. Positions are open for almost any type of scientist and technician. But because of its preponderant hardware commitments, *engineers* are the human prizes most wanted. Any specialty from Automation Engineering to Zodiac Astrogration Technology can be utilized. NAA wants YOU, if you want to be one of the "men behind the rocket engines" that drive our craft into the big dark beyond earth.

DOUGLAS AIRCRAFT CO., INC.

3000 Ocean Park Blvd., Santa Monica, California

Profile

Founded 1920. Went heavily into military missiles with Thor IRBM, 1955 on. Exceeded $1 billion gross sales in 1960. Slipback below that mark in 1961 due mainly to company reorganization. Big Saturn IV and IV-B stages, in Apollo Program, puts firm in strong position to 1970 (Apollo contracts form 8.8% of total business in 1963; percentage will climb yearly). Total personnel 48,-000; 37% technical. Total missiles made to date: over 40,000.

Astronautics Feats

—R&D of Thor IRBM under USAF crash program, completed in record-breaking 13 months.

—Thor missile reached 100th firing in October, 1961, both as missile and space booster, for exceptional reliability record of 87%.

—Thor booster in varied combinations—Thor-Able, Thor-Delta, Thor-Ablestar, Thor-Agena—is "workhorse of space," launching 70% of all USA space vehicles.

—Thor-boosted launchers in 1963 passed mark of 75 total satellites and interplanetary probes listed as successfully propelled into space.

Company Characteristics

Aggressively seeking larger share of space budget with its Missile & Space Systems Division growing most rapidly. Researches into advanced astronautics, partly NASA/USAF funded, include nuclear power systems, astronavigation, deep-space exploration, bionics, human engineering, solar-powered spacecraft (via LASER technique). Inducements to scientists, engineers and technicians:

scholarships toward higher degrees on full-time or part-time school-ing basis; on-the-job training formerly limited, now being in-creased.

Divisions and Subsidiaries

Missiles & Space Vehicles Division, same address as parent cor-poration. Production of military missiles (Thor, Genie, Nike-Zeus, Skybolt) and spacecraft (Thor-Delta, Saturn upper stages, Big Shot rigidized ComSat). Many applied-research facilities, such as their Aerophysics Laboratory, conduct large variety of ex-otic investigations into future astronautics. Employees, 21,000; 40% technical.

Aircraft Division, 3855 Lakeview Blvd., Long Beach, California. Production of military planes (USN Skyhawk series, etc.) and commercial jetliners (Douglas DC-8).

Tulsa Division, 2000 N. Memorial Drive, Tulsa, Oklahoma. Aircraft reconversion work and some phases of missile systems.

Charlotte Division, 1820 Statesville Rd., Charlotte, North Caro-lina. Other military missiles produced: Nike-Hercules, Nike-Ajax, Honest John, Rockaire.

Aircomb Division, 3000 Ocean Park Blvd., Santa Monica, Cali-fornia. Manufactures nose-cone and structural casing materials for military missiles.

Astropower, Inc., 2968 Randolph Ave., Costa Mesa, California. A separate subsidiary doing R&D into better liquid rocket fuels, sea-launch for space boosters, fuel-cells for spacecraft, and other advanced projects.

Main Contracts (Missiles and Space Only)

Current (Prime). Thor-Delta, full 3-stage launcher for NASA; Thor IRBM for USAF; Thor boosters for other launchers, NASA/USAF; Honest John, U. S. Army; Genie, air-to-air for USAF; Rockaire, JATO/RATO unit; Big Shot, Echo follow-on, 135-foot passive radio-relay balloon satellite; Saturn IV, 2nd stage, for NASA Saturn C-1; Saturn IV-B, 2nd stage, for NASA Saturn C-1B.

(*Sub-Prime*). Nike-Zeus AICBM, missile parts and GSE, for Army; Nike-Ajax, for Army; Nike-Hercules, for Army; Sidewinder, USN/USAF; Zuni, USN.

Future (*Probable*). Thor space boosters for NASA/USAF (indefinitely); Thor IRBMs (will phase out when Pershing solid-fuel missile is operational, circa 1964-65); Nike-Zeus AICBM (fate formerly uncertain, but DOD now recommends continued R&D); Nike-Hercules, continuation as standard anti-aircraft missile throughout nation; Saturn upper stages for C-1 extended to C-5 at increasing yearly rate to beyond-billion-dollar totals by 1970; Nuclear propulsion, fuel-cell technology and manned space-flight vehicles likely areas in which Douglas may win future contracts.

Technical Help Wanted

Expansion plans with new Saturn contracts include filling posts in 106 specialized space-technology fields, spreading through 37 scientific/engineering disciplines. Main needs are: *Engineers*: Aeronautical; Astronautical (all types); Chemical; Electrical; Electronics (all types); Nuclear; Chemo-nuclear; Metallurgical. *Scientists* (*applied and research*): Astro-mathematics; Astro-physics; Bionics & Bio-astronautics; Geophysics; Micro-biology; Nuclear physics; Psycho-physics; Astro-psychology.

Summary

With little doubt, Douglas will remain among any Big Ten listing of industries in military missiles plus space technology. Its entry into the giant Apollo Program can mean spectacular company growth until 1970, if not beyond. Its firm-sponsored research activities after a slow start are rapidly moving up into the high-level class, with an estimated 1000 to 3000 quality researchers doing notable space-oriented work. Wide open opportunities for scientists, engineers and technicians in almost any category.

THE BOEING COMPANY

Box 3707, Seattle 24, Washington

Profile

Founded 1916. Called Boeing Airplane Co. until 1959. Leader in large jetcraft, both military (B-52 bomber, KC-135 tanker) and civilian (707 and 720 airliner). Among top aerospace firms in gross sales (around $2 billion mark now) for many years. Total personnel: 95,000.

Astronautics Feats

—Developed Minuteman, first solid-fueled ICBM, which will replace all liquid-fueled Atlas and Titan missiles within 5 years.

—Conceived Dyna-Soar, first manned maneuverable spacecraft for orbital missions.

Company Characteristics

Specializes in massive contractual projects, typified by mass production of B-52 jet bombers and Minuteman missiles. Income mainly from DOD/USAF rather than NASA, but with recent Saturn S-1B award (bottom booster of Saturn C-5) is now orienting toward space. Project PARSECS (company feasibility study) is a complex of proposed manned space stations and deep-space outposts, which may follow successful Dyna-Soar experiments. The corporation has strong management and administrative teams of top technical calibre and constantly seeks more.

Divisions and Subsidiaries

Boeing Associated Products, same address as parent corporation headquarters. Plant producing non-aerospace diversified products in electronics and structural materials.

Transport Division, Box 707, Renton, Washington. Produces USAF jet tankers and commercial jetliners.

Aerospace Division, Box 3707, Seattle 24, Washington. Production of Minuteman, Dyna-Soar, Saturn S-1B, interceptor missiles, hydrofoil (water-skimming) craft.

Military Aircraft Systems Division, Box 3999, Seattle 24. Production of B-52 Stratofortresses (phased out in 1962).

Industrial Products Division, Box 3955, Seattle 24, Washington. Diversified products.

Vertol Division, Woodland Ave. & P.R.R., Morton, Pennsylvania. R&D of experimental VTOL (Vertical Take-Off and Landing) jetcraft.

Main Contracts (Excluding Aircraft)

Current (Prime). S-1B booster for Saturn C-5 with 7.5 million pounds thrust from five F-1 engines (NASA); Minuteman, solid-fueled ICBM, operational in 1962, will reach total of over 800 in silos by 1968 (USAF); Bomarc, A & B models (ramjet and solid-fueled) surface-to-air missiles (USAF); X-20 (Dyna-Soar), manned maneuverable orbital craft, rocket-boosted into space (USAF).

(Sub-Prime). Minuteman. GSE and transport vehicles (USAF).

Future (Probable). ASP (AeroSpace Plane), follow-on of Dyna-Soar, with ground take-off capability (R&D up to 1970); Extension of Saturn C-5 booster program past 1970, running into multi-billion-dollar totals; Nova, booster or upper stage, starting 1964-65; Some phase of manned orbital space stations (NASA and USAF reviewing bids and proposals of Boeing and other competitors).

Technical Help Wanted

Missile engineers of all kinds are in demand, plus aeronautical/astronautical flight specialists. Scientists for advanced space systems research are also being sought. Best opportunities are for technicians in any class as well as managerial talent (who are winnowed out of each new young technical crop and offered company train-ing).

Summary

Boeing has stayed in a strong position by seeking out the big long-term contracts. When its lucrative B-52 contracts ended in 1962, it was already deep in mounting production of the Minuteman, which will be America's major strategic missile at least through this decade. Using foresight in the space program, Boeing has worked up engineering studies of a vast manned-vehicle complex, includ-ing such unique items as a "counter-moon" (opposite the moon in same orbit), cis-Mars and cis-Venus "space garages" for interplane-tary ships, and "out-of-ecliptic" craft to venture away from the solar system's plane into the fringes of interstellar space. If the Dyna-Soar project (starting flights in 1964) goes well (and isn't cancelled), Boeing will be in a good position to obtain major manned-space-vehicle contracts now being contemplated by both NASA and the USAF.

CHRYSLER CORPORATION

341 Massachusetts, Highland Park, Michigan

Profile

Well-known for its automotive history, Chrysler was first in that industry to obtain big missile contracts, starting with Redstone/Ju-piter IRBM for Army/USAF in 1950, which led to Jupiter-C and Juno space launchers for NASA from 1958 on. Spurred by recent Saturn booster contract, new Space Division was formed in early 1962. Major business has been, and will be, motorcars, but per-centage of astronautics gross and personnel is increasing steadily. Total employees: 80,000.

Astronautics Feats

—First USA satellite (Explorer I, January 31, 1958) orbited by Jupiter-C launch vehicle.

—First moon flyby probe (Pioneer IV, March 3, 1959) shot to escape velocity by Juno-2 launcher.

—Eight early satellites propelled into space by Jupiter-C and Juno-2 vehicles before other boosters (Thor and Atlas) reached performance stage.

—Jupiter/Juno boosters lofted all space vehicles responsible for great discovery of Van Allen Belts (Explorers I, III and IV, Pioneers III and IV).

Company Characteristics

Though failing to aggressively follow up its early military-missile experience for several years, Chrysler in 1960 swung back in with rocket electronics, then won the first Saturn contract for the C-1's bottom booster. Later came the booster award for Saturn C-1B (3-stage instead of 2-stage Saturn C-1 vehicle). With the formation of its new Space Division, Chrysler is now geared to branch out into other phases of space work.

Divisions and Subsidiaries (Missiles and Space Only)

Airtemp Division, 1610 Webster St., Dayton, Ohio. Aerospace instrumentation.

Amplex Division, Box 2718, Detroit 31, Michigan. Electronics.

Missile Division, Box 2628, Detroit 31, Michigan. Former Redstone and Jupiter works, now maintenance only for emplaced operational missiles.

Space Division, Box 26018, New Orleans 5, Louisiana. Production of Saturn C-1 and C-1B boosters (at NASA's Michoud Plant).

Main Contracts (Missiles/Space Only)

Current. Saturn S-1 first-stage boosters for NASA (30 on order); Saturn booster for C-1B (8 on order); Sub-prime and sub-contracting work in miscellaneous instrumentation, electronics and missile parts.

Future (Probable).   Saturn C-1 and C-1B bottom boosters, to follow first orders, may total over 100; Sub-prime contracts for Saturn C-5 boosters; Possible sub-contracting on nuclear Saturn and Nova, about 1964-65; Expansion into other phases of astronautics: propulsion, guidance, manned flight systems, lunar rover (surface) vehicles, radio/telemetry.

Technical Help Wanted

Every type of engineer and technician for expanding Saturn booster program, now getting under way and reaching a peak by 1964-65. Applied science researchers for advanced space technology up to and beyond lunar program.

Summary

As with General Motors and Ford, Chrysler is committed to maximum efforts in the astronautics industry (without, of course, minimizing its well-established motorcar business). The Saturn booster contracts represent a definite jump over its automotive rivals, at present. As yet comparatively low, dollar-volume astronautics grosses for Chrysler will increase significantly, plus the "guaranteed" future income that the Apollo moon project bestows.

10 : ASTRONAUTICS ACES

(*Profile and résumé of other aerospace primes . . . Martin-Marietta . . . General Electric MSD . . . Lockheed . . . General Dynamics . . . Aerojet-General . . . United Aircraft . . . McDonnell . . . Thiokol . . . Thompson Ramo Wooldridge . . . Radio Corp. of America.*)

The Big Four in the Apollo moon program are not necessarily the foremost firms in the total space picture. Others are also major prime contractors handling unmanned satellites, probes and non-lunar manned systems.

Then, too, we must remember that at present military rocketry plus jetcraft account for the greatest proportion of "aerospace" funding. In FY-1962 no less than $25,900,000,000 (almost $26 *billion*) of DOD's total defense budget of $50 billion went into jets and missiles. Out of the $25.9 billion, about $13 billion was allocated to missiles and space, excluding jets: $1,600,000,000 for space projects, $4,200,000,000 for R&D of military missiles, and $7,200,000,000 for procurement and development of rocket weaponry.

Hence the "aerospace" portion of any firm's gross sales is preponderantly DOD contracting, and the larger part of this sum is in military jetcraft and ballistic missiles rather than in space itself. However, the top hundred "aerospace" firms also represent the top hundred "space" firms, in that both fields deal basically with the same thing—rocket-powered machines and their payloads.

The following group might be called a "Top Technological Team" whose services—apart from their DOD/missile work—are vital to our national space program.

MARTIN-MARIETTA CORPORATION

Friendship International Airport 40, Maryland

Profile

Founded by Glenn L. Martin, famed air pioneer, in 1912 (as The Martin Co.). Built 12,000 planes in a half-century. Entered missile field in 1952. Only aircraft firm to drop making planes (1960) and switch entirely to missiles. Merged with Marietta in 1961. Total personnel: 40,000. Annual dollar-volume gross exceeds $1 billion.

Astronautics Feats

Produced Viking, first all-American designed big-thrust liquid-fueled rocket (superseding the V-2). Prime for Vanguard Project, America's first space program, which orbited three satellites, including now-famed Vanguard I, 6-inch "baby" satellite that made space history by revealing earth's pearlike shape. Vanguard upper-stage assembly, converted to "Able" stage atop Thor booster, later orbited long series of satellites.

Main Contracts (Prime)

Military missiles, including USAF's Titan ICBM, USN's Bullpup, Army's Pershing. For space projects, currently developing Titan-2 and Titan-3 as man-rated boosters for Gemini and Dyna-Soar.

Divisions and Subsidiaries

Aerospace Division, headquartered at Friendship International Airport, Maryland.

Space Systems Division, Baltimore 3, Maryland. Largest division (15,000 employees), produces military missiles (except Titan).

Denver Division, Box 179, Denver 1, Colorado. Titan production plants, plus advanced space researches; 12,000 employees.

Orlando Division, Box 5837, Orlando, Florida. Electronics; 8000 employees.

Nuclear Division, Baltimore 3, Maryland. Design of nuclear SNAP units. Staff of 400 experts.

Electronic Systems & Products Division, Baltimore 3, Maryland. Undersea and radar apparatus.

RIAS Division, 7212 Bellona Ave., Ruxton, Maryland. Research Institute for Advanced Studies, founded 1955; 50% federally subsidized; top-rated pure research laboratory.

Canaveral Division, Cocoa Beach, Florida. 200-man technician team aiding in Martin missile launchings.

Technical Help Needed

All types of engineers, but particularly in the areas of thermo-flow, stabilization, flight dynamics, fuel technology and reliability. Also engineering scientists in non-linear math, nuclear/electric propulsion, isotopic power, astro-data computers, extraterrestrial metallurgy, lunar surface vehicles, space bionics.

Prospects of Corporation

The firm's philosophy is summed up in a message to employees from management: "The future? Men of Martin-Marietta cannot read it—but they are helping to *determine its shape*." Their advanced study of the "Martin Moon Colony" is typical of their strong bid to play a significant role in our future space program.

GENERAL ELECTRIC CO. (Missiles and Space Only)

570 Lexington Ave., New York 22, New York

Profile

History as utility well known (4th largest U.S. corporation). Entered military missile field early, before 1950. Several divisions and

departments today derive major gross from aerospace field, totaling
$1.2 billion today, or 25% of GE's full income. Company is geared
to heavy future commitments in space. Present aerospace technical
personnel is 19,000, including 5000 engineers and scientists hand-
picked for ability.

Astronautics Feats

Pioneered first re-entry nose-cone of Jupiter IRBM, in 1958. Developed Hermes (among earliest USA missiles). Designed first
camera to take both black-and-white and color movies of earth
from high-altitude rockets. Built first large-scale Space Environment Chamber able to swallow full-sized satellites and vehicles
whole.

Main Contracts

For USAF, nose-cone prime for more than 50% of all military missiles; Agena/Discoverer re-entry system. NASA prime for Nimbus
Weath-Sat; Apollo Craft Reliability Management. Sub-prime for
Samos, Atlas guidance, Polaris firing-control, missile payload shielding for re-entry (Titan, Thor, Skybolt). Subcontracts in a dozen
other missiles and space projects. Fuel-cell system (on board power)
for Gemini craft.

Divisions and Departments (Missile/Space-Oriented Only)

Defense Electronics Division (and *Defense Systems Department*), Electronics Park, Syracuse, New York. Guidance, tracking
and communications systems for rockets.

Missile and Space Division and *Valley Forge Space Technology
Center*, Box 8555, Philadelphia 1, Pennsylvania. Production of
nose-cones, space vehicle systems, missile airframes, other aerospace
paraphernalia. Research into satellite power techniques, exotic
communications, interplanetary exploration, most other astronautics ventures.

Flight propulsion Division (West Lynn 3, Massachusetts); *Heavy
Military Electronics Department* (Court St., Syracuse, New York);

Light Military Electronics Department (French Road, Utica, New York); and *Direct Energy Conversion Operation Department* (West Lynn 3, Massachusetts) are other parts of the GE empire partially but not extensively involved with astronautics work.

Technical Help Needed

Constant recruitment of engineers, scientists and technicians in plasma dynamics, hydromagnetics, hypersonics, spacecraft flight dynamics, guidance and communications, ecology in bio-astronautics, zero-g space tools and space rescue systems. GE's Service Training Unit provides excellent on-the-job training, and their Scholarship Student Aid Program is well financed.

Prospects of Corporation

The giant (900,000 square feet) new Valley Forge facility of MSD was specifically designed to put GE into the front-running in our space program. The Center's lavish laboratoryware is only outclassed by its scintillating brainware, whose top-drawer quality is typified by Dr. George Arthur, Dr. Leo Steg and Dandridge Cole. GE's research teams are going at hypersonic pace into frontier technologies—planet-to-planet spaceplanes, orbit ferry service, zero-g spacesuits, satellite repair, space-station logistics, lunar camp powerplants—yet without ignoring more immediate space plums such as production of nose-cones. GE's astronautics expansion will undoubtedly see no slacking off in the predictable future and lifetime careers beckon for across-the-board technical manpower.

LOCKHEED AIRCRAFT CORPORATION

2555 N. Hollywood Way, Burbank, California

Profile

Founded 1912. Famed for planes, such as F-104 Starfighter and notorious U-2. Became well diversified (ships, nucleonics, heavy construction) and topped $1 billion gross from 1959 on. Entered astronautics early with rocket hardware, which passed its aircraft business from 1961 on. Total employees: 95,000.

Astronautics Feats

Agena upper-stage placed USA satellite (Discoverer) in first polar orbit. Agena-borne Midas and Samos were pioneer ReconSats. Thor-Agena launch combination has lofted record-breaking number (over 50) of USA satellites and probes. Thor-Agena-B sent Mariner-2 on recent Venus flyby mission for first contact with another planet. First 5-satellite launch (with Agena-D, in Feb., 1963).

Main Contracts

Agena-B and D vehicles for USAF will continue increasing yearly (39 in 1962) with Thor and Atlas boosters, plus all Discoverer, Midas and Samos satellite systems. Prime for Polaris missile of nuclear submarines (USN) and for Rover nuclear rocket R&D (AEC/NASA). Sub-prime for propulsion unit of Apollo craft in moon program.

Divisions

Lockheed-California Co., 2555 N. Hollywood Way, Burbank, California. Production of military and civilian aircraft.

Lockheed-Georgia Co., 86 S. Cobb Dr., Marietta, Georgia. Various aerospace components.

Lockheed Missiles & Space Co., Box 504, Sunnyvale, California. Production of Agena space vehicles and Polaris missiles; also advanced astronautics research.

Lockheed Propulsion Co. (formerly Grand Central Rocket Co.), Box 111, Redlands, California. Solid-fuel rocket systems.

Lockheed Electronics Co., U. S. Highway 22, Plainfield, New Jersey. Diversified electronics products.

Technical Help Wanted

Engineers in perpetual demand in 27 categories, technicians in all fields, scientists for applied research and advanced astronautics theory. Solid-fuel specialists wanted for "permanent" employment, also Agena staffs. (Note: Much of Lockheed's work (ReconSats, Polaris, etc.) is under classified regulations; employees must be cleared under DOD security requirements.)

Prospects of Corporation

The firm's DOD contracts for military missiles and space vehicles make up most of their aerospace contracts, and will extend into an indefinite future. Contracts for NASA are comparatively minor as yet (under $10 million) but with sub-prime and subcontract portions of the Rover Project and Apollo Program, Lockheed's nonmilitary efforts will perforce increase. Around 1964-65, when Rover nuclear-powered rockets go into second-phase R&D, funding will increase enormously. Lockheed jobs in general can be called "long-term" with low chance of its aerospace endeavors meeting "sudden death" cancellations.

GENERAL DYNAMICS CORPORATION

1 Rockefeller Plaza, New York 20, New York

Profile

Famed for a quarter-century for its planes (F-106 Delta Dart and Convair 990 as of today), firm gained equal fame for its Atlas ICBM missile, which became America's biggest space booster in the "Space Race" after 1957. Derives major income from rocketry today, with large contracts both for DOD/USAF and NASA projects. Passed into $2 billion sales class in 1961, but low profits required a reorganization of management and divisions that has now rejuvenated the firm. Future space commitments, as well as present, assure GD a strong place in the astronautics industry through this decade and beyond. Total employment: 95,000.

Astronautics Feats

First "talking satellite" (Project Score, December, 1958) orbited within Atlas nose-cone. First USA astronauts in multi-orbit space-flights via Atlas booster. Atlas-Agena payload first to strike the moon (Ranger-4, April 24, 1962). First multi-ton USA payloads were Atlas-borne (Midas and Samos, 5000 pounds). First hydrogen-fueled second-stage vehicle (Centaur). In military missiles, pioneered bazooka-type military rocket (Redeye) and "cargo" missile (Lobber).

Main Contracts

Prime for Atlas series (A to F) ICBM's and space boosters; for Atlas-Centaur launch vehicle; for Nova (co-prime is Martin); for ARENTS (Advanced Research Environment Test Satellite) in Project Vela Hotel (detection of nuclear tests in space); with co-prime General Atomics, will produce Orion nuclear-pulse propulsion (via small atomic explosions); prime for USN Tartar and Terrier, Army Redeye, Lobber and Mauler. Subcontracts in Apollo Project; prime for Little Joe II launcher to test boilerplate Apollo craft.

Divisions

General Dynamics/Astronautics, 5001 Kearny Villa Rd., Box 1128, San Diego 12, California. Production of Atlas, Centaur, ARENTS and related electronic components.

General Dynamics/Convair, 3302 Pacific Highway, San Diego 12, California. Atlas booster airframe parts, GSE, varied aerospace instrumentation.

General Dynamics/Fort Worth, Box 748, Fort Worth 1, Texas. Subcontracting for aerospace items.

General Dynamics/Pomona, 1675 W. Fifth St., Box 1011, Pomona, California. Military missiles.

Technical Help Needed

Military missile engineers and technicians are high on the list, but also a growing number of engineering scientists for tomorrow's technology involving advanced propulsion and its concomitant techniques. Technically trained management and administrative personnel are sought constantly because, like Boeing, GD deals in massive projects requiring large pools of brainpower.

Prospects of Corporation

Though the great Atlas ICBM project of the past 10 years is now approaching phase-out, Atlas boosters will still be used for space work by both the USAF and NASA. The Centaur Program, ailing for two years, is now regalvanized to take up slack, and such future projects as Nova and Orion will eventually pull in major funding. Other company-initiated studies (under the brilliant Krafft Ehricke) include complete interplanetary expeditions and space-station systems that may land GD multi-billion dollar contracts before and after 1970.

Through its Astronautics division, the firm will unquestionably be in the space forefront and play an important astronautics role.

AEROJET-GENERAL CORPORATION

Azusa, California

Profile

A subsidiary of the General Tire & Rubber Co., handling its missile/space contracts. Founded 1941 as Aerojet Engineering Co.; acquired by GT&R in 1944 under the new name. A-G is one of the few "all-aerospace" firms, born of the missile/space age and gathering its entire sales in astronautics. It is the foremost producer of solid-fuel rockets and second only to Rocketdyne in liquid-fuel engines. Its growth has been phenomenal, from 375 employees and a gross of $2,800,000 in 1946 to 40,000 personnel producing about $700 million today. No less than 10,000 of its total labor force hold science and engineering degrees. It has built more new plants and astronautics laboratories in the past five years than any other aerospace firm.

Astronautics Feats

Produced first operational upper-atmosphere research rocket in the Aerobee. First JATO and RATO devices. First big booster solid-fuel engine to be designed and tested (500,000 pounds thrust). First restartable upper-stage engine to correct satellite orbits (Able-star). First orbiting of two satellites at once (also three) in Transit series (Ablestar has unique distinction of placing 10 satellites in space with 5 launchings).

Main Contracts

Prime for NERVA (Nuclear Engine for Rocket Vehicle Applications); for Aerobee series, Spaerobee, Astrobee and Aerojet Jr. research rockets. Sub-prime in propulsion for Titan engines, Atlas-Able, Thor-Able and Thor-Ablestar launchers; Bomarc booster; solid-fuel engines for tactical military missiles (Sparrow-3, Falcon series, Tartar, Skybolt, Genie, Hawk, Regulus-II), and for giant Polaris and Minuteman; upper-stages of solid-fuel space boosters (Scout, Blue Scout and Blue Scout Jr.). In future vehicles, A-G engines will go into stage of the Nova (M-1 hydrogen-fueled unit with 1.4 megapounds thrust). In the moon program, sub-prime propulsion systems for the Apollo spacecraft and Saturn launch vehicles. For the USAF, Dyna-Soar propulsion and Big Solids Program for Titan-3 booster, which NASA will also use.

Divisions and Departments

Azusa Plant, Box 296, Azusa, California. Solid/liquid propellant rocket engines, test facilities and production of solid fuels.

Aerojet-Delft Corp., Engineers Hill, Plainview, Long Island, New York. Engineering studies and systems.

Aerojet-General Nucleonics, Box 77, San Ramon, California. Electronics and instrumentation.

Atlantic Division, Box 460, Frederick, Maryland. Automation systems.

Aetron Plant, 410 N. Citrus, Covina, California. Engineering and support equipment, electronics and control devices.

Downey Plant, 11711 Woodruff Ave., Downey, California. Jet drone vehicles and military missiles.

Sacramento Division, Box 1947, Sacramento, California. Two plants (Liquid Rocket and Solid Rocket) doing advanced R&D on big-thrust engines.

Space-General Corp., 9200 E. Flair Dr., El Monte, California. Actually a separate subsidiary of GT&R, launched in 1961 and growing rapidly. It is taking over much of production of Aerobee rocket, Ablestar engine and advanced space technology studies.

Technical Help Wanted

Engineers and technicians specifically oriented toward missiles and space vehicles have fine opportunities here, particularly propulsion specialists (on-the-job training offered). Nuclear scientists, astrophysicists and other research personnel are rapidly being added to the roster, with want-lists open for years ahead. Rapid advancement for recruits with special ability is a company policy.

Prospects of Corporation

The firm's rising gross-income curve indicates it will continue overhauling others and probably reach top-levels of space contracting within this decade. It is advancing in all areas of rocket-propulsion —solid, liquid, nuclear, electric, exotic. Key subcontracting for propulsion units in future manned spacecraft assures mushrooming projects in the Apollo Program and beyond. Its straight-line Space Age business orientation is superbly directed by a foresighted management team headed by President Dan A. Kimball.

UNITED AIRCRAFT CORPORATION

400 Main St., East Hartford 8, Connecticut

Profile

Founded 1925. Gross passed billion-dollar mark in 1961. Largest producer of jet engines, aircraft accessories and helicopters, which with non-aerial products make up some 90% of total sales. UAC is thus the "latest and least" of the aircraft makers to enter aerospace. Two of its divisions pioneered in high-energy (hydrogen) rocket-fuel technology and solid rockets for future giant booster development, which promise to have a far greater future share in the astronautics industry. Total employees: 75,000.

Astronautics Feats

No outstanding "space firsts" can be attributed directly to UAC hardware or technology, because of its late start in the Space Age. However, its RL-10 rocket engine is the first to burn hydrogen fuel (15,000 pounds thrust), and it has obtained DOD contracts for R&D of giant solid-fuel boosters (120 inches to 256 inches in diameter) to be strapped onto the Titan-2 to make a powerful Titan-3 booster. This vehicle will share the Saturn's workload from 1964 on, and may loft many Apollo 3-man capsules. UAC also introduced the first feasible concepts of "segmented" solid-fuel modules that could be piled up like blocks to make super-boosters of any desired thrust in multi-megapounds.

Main Contracts

RL-10 engines for Centaur, and for Saturn S-IV second-stage; solid-fuel Titan booster auxiliaries; inboard power system for Apollo spacecraft; ramjet engine for USAF Hound Dog missile; light-weight nuclear reactor for SNAP Program; environment control system for lunar LEM.

Divisions

UAC Headquarters, East Hartford, Connecticut. Parent plant for aircraft production.

Pratt & Whitney Aircraft Division, 400 Main St., East Hartford 8, Connecticut. Jet engines and RL-10 rocket motors.

Hamilton Standard Division, Windsor Locks, Connecticut. Propellers, aircraft components, aviation instrumentation.

Sikorsky Aircraft Division, Stratford, Connecticut. Large-scale production of helicopters for all the armed services.

Norden Division, Helen St., Norwalk, Connecticut. Guidance, control and other electronics devices for aircraft and missiles.

Florida Research & Development Center, Box 2691, West Palm Beach, Florida. Advanced research, including LASER's, plasma physics, astro-navigation.

United Technology Corporation, Box 358, Sunnyvale, California. Design and testing of high-thrust solid-fuel boosters, and research into other propulsion concepts.

UAC Systems Center, Windsor Locks, Connecticut. Specialized studies in engineering systems to coordinate big missile/space projects.

Technical Help Needed

Outside of aeronautical engineers and related fields, aerospace brainpower is being gathered in their new Space Age centers. Specialists in rocket propulsion, solid or liquid, will find many openings; also any area of electronics, physics, math, biology and atomics related to astronautics. Watch for a rapid rise in aerospace employment in the next few years. Liberal scholarship (undergrad and grad) aids and on-the-job training programs for "B" grade or better students are offered at all UAC divisions. Training projects for technicians have just begun with rapidly growing enrollments.

Prospects of Corporation

Other business aside, UAC's richly varied diversification, plus large-team engineering experience, will undoubtedly bring in the choice contract plums that have been falling to other aircraft companies who have swung heavily toward rocketry. Solid-fuel boosters in the Nova class, advanced hydrogen-burning engines and propulsion aspects of the Titan-3 Program are fields in which UAC will reap new missile/space business.

MCDONNELL AIRCRAFT CORPORATION

Lambert-St. Louis Municipal Airport, Box 516, St. Louis 66, Missouri

Profile

Founded 1939. Major sales in military aircraft (Phantom, Voodoo, Demon). Entered astronautics field in 1959 with "glamorous" Mercury capsule contract, leading to follow-on Gemini craft. Diversification into electronics and automated servomechanisms is leading to further subcontracting in aerospace products. Total employees: 33,000.

Astronautics Feats

Beginning with Shepard in May, 1961, all spaceflights by American astronauts occurred in Mercury capsules. And with the ending of the Mercury Program, Gemini craft will take over all further orbital feats by 2-man teams.

Main Contracts

With phase-out of Mercury capsules, Gemini construction is in full swing on an open-end contract (number of vehicles not yet stated). In military missiles, prime for USAF target-drone Quail, sub-prime in propulsion for USN Talos (phasing out), USN Typhon R&D (anti-missile missile) and Army TOW studies (anti-tank missile).

Divisions

McDonnell has none as yet. Their entire plant and production facilities are located at St. Louis.

Technical Help Needed

A variety of technologists in metallurgy, life-support, electronic servo-controls and other systems relating to the Gemini manned craft, including specialists in orbital rendezvous mechanics, zero-g dynamics and astro-guidance.

Prospects of Corporation

Despite its comparatively limited projects in astronautics so far, McDonnell may be a "sleeper" destined to gain a major role. This is foreshadowed by USAF participation in NASA's Gemini Program, which may increase enormously in order to develop orbit-rendezvous techniques. Also, the firm's engineering studies of advanced space systems—lunar logistics, interplanetary re-entry techniques, earth-to-orbit ferry craft—are aimed at substantially increasing its role in this nation's space effort. The invaluable experience of McDonnell's engineering teams in designing the very first USA manned space vehicles is a plus value that NASA will certainly continue to utilize in future spaceflight endeavors.

THIOKOL CHEMICAL CORPORATION

Box 27, Bristol, Pennsylvania

Profile

Founded 1929. Synthesizing a chemical called *thiokol* gave this firm three things—its name, a vast new market in artificial rubber, and (remarkably enough) an important new solid fuel for rockets. Thiokol-based fuels (mixed with oxidizers) made the company second only to Aerojet-General in the field of military missiles. In 1958, Thiokol acquired the pioneer of liquid-fuel rocket technology in America—Reaction Motors Incorporated—which had produced the first commercial rocket engine in 1941. From then on Thiokol/RMI made Space Age history time and again by powering the early Lark and Viking missiles, and later, the Skyrocket, X-1, X-1A and X-2 rocketplanes that shattered all speed records and broke the sound barrier in 1947. RMI also began developing the XLR-99 engine, which was finished after affiliation with Thiokol, to become the powerplant of the famed X-15. Thus, through the Space Age since 1957, Thiokol has made enormous strides both in astronautics technology and gross sales. Total employment: 38,000.

Astronautics Feats

Opened entire new field of high-speed reliable, mass-produced propulsion units for missiles by introducing the rubber/asphalt/plastic-based solid fuels. Perfected RMI's throttleable engine with which the X-15 performed as the world's first piloted spacecraft, achieving altitudes of 60 miles and hypersonic (5 times sound) speeds beyond 4000 mph by the end of 1962.

Main Contracts

Sub-prime for propulsion systems of military missiles (R&D or operational status)—Minuteman ICBM, Lacrosse, Matador, Mace, Pershing, Sergeant, Bullpup, Nike-Ajax, Nike-Hercules, Nike-Zeus, Subroc, Falcon series, Bomarc-B, Honest John, Sparrow-3. NASA sub-prime propulsion contracts for Apollo spacecraft, Scout and USAF Blue Scout series. Original X-15 contract for three engines optioned to continue into advanced X-15 craft for training USAF space pilots in Dyna-Soar program.

Divisions

Separate plants with more or less interrelated production facilities in solid-fuel rocket engines, liquid-fueled propulsion and other missile paraphernalia are scattered widely: *Rocket Operations Center* (3340 Airport Road, Ogden, Utah), *Wasatch Division* (Box 524, Brigham City, Utah), *Redstone Division* (Huntsville,

Alabama), *Longhorn Division* (Marshall, Texas), *Elkton Division* (Box 241, Elkton, Maryland), *Panelyte Division* (N. Enterprise Ave., Trenton, New Jersey), *Trenton Division* (780 Clinton Ave., Trenton 8, New Jersey), *Moss Point Division* (Box 517, Moss Point, Mississippi), *Logan Works* (Peterson Bldg., 2503 N. Main St., Logan, Utah).

Reaction Motors Division, Denville, New Jersey. R&D of liquid-fuel rocket engines.

Reaction Motors Production Plant, Box 27, Bristol, Pennsylvania. Operational propulsion systems.

Shawnee Industries, Inc. (subsidiary), Box 549, Shawnee, Oklahoma. Missile components, plastics.

Technical Help Wanted

Missile engineers and technicians in all categories, aerodynamic and astrodynamic physicists, chemical and metallurgical research technologists, science theorists in any of the physical sciences (including astronomy, mathematics and applied biology). Thiokol's rapid growth pattern indicates a dire need for new brainpower into the indefinite future.

Prospects of Corporation

Geared almost wholly to the Space Age (like Aerojet-General), Thiokol is in an excellent and in fact unshakable position to ride the crest of the astronautics tidal wave, probably to the end of this century and beyond. As fast as military contracts phase out with obsolescent missiles, new ones come in. In the space program, Thiokol is sure to be a sub-prime for any firm doing future giant solid-fuel boosters. The X-15 program is not ending but just beginning, as the USAF completes plans to use this "trainer" for space pilots in the Dyna-Soar, Gemini and advanced man-in-space projects to come. Astronautics, one might put it, is Thiokol's middle name.

THOMPSON RAMO WOOLDRIDGE INC.

23555 Euclid Ave., Cleveland, 17, Ohio (Eastern HQ). Western HQ: 8433 Fallbrook Ave., Canoga Park, California

Profile

A unique firm dealing largely in "engineering system services." Its famed subsidiary, Space Technology Laboratories Inc. (before 1958 the Guided Missile Research Division) became a special "prime prime" for the USAF, or the "technical overseer" for the entire R&D systems behind the Thor, Atlas, Titan and Minuteman missiles. Other divisions and subsidiaries of TRW have become active in the aerospace field, from which it today derives 50% of its gross sales. Its engineering teams and space research staffs are unmatched for their ability to integrate projects involving complex hardware. Employees: 33,000.

Astronautics Feats

Indirectly, TRW's STL shares the honors for many early USA space feats, including adaptation of the Thor and Atlas into space boosters that lofted dozens of satellites or probes, and sent astronauts into orbit. Specifically, STL produced the first "paddlewheel" satellite (Explorer VI, April, 1959) for outer space studies (apogee, 26,000 miles) and the sensational Pioneer V (March, 1960), which sent the first telemetry signals from interplanetary distances (22.5 million miles).

Main Contracts

Prime-prime program manager for Minuteman production, Titan-2 (both ICBM and manned booster) and Titan-3 (solid-boosted space launcher to come). Also for various top-secret and classified USAF space-weapons systems, typified by BAMBI (Ballistic Anti-Missile Boost Intercept). Prime for NASA's series of earth-study satellites (OGO, EGO, POGO). Sub-prime with major projects such as NAA's Apollo, Boeing's Dyna-Soar, Syncom communications satellite of Hughes, and Nike-Zeus of Western Electric. An important specialty (via STL) is supplying the vital programming sequences of computer/guidance systems for space-flights—typified by the Mariner-2 Venus probe, the astronauts' orbital flights and various earth satellites.

Divisions and Subsidiaries

One group of these is connected only indirectly, as suppliers of parts or minor engineering aids, with the aerospace field—Automotive Group, Thompson Products Replacement Division, Electro-Mechanical Group, Jet Division, Dage Television Division, Magna Corp., Radio Industries, Inc. and others.

Electronics Group, 8433 Fallbrook Ave., Canoga Park, California. Control and guidance devices for spacecraft.

Intellectronics Division, Same address. Computer systems for rocketry, plus research into "thinking" machines. ("Intellectronics" is derived from "intellect" and "electronics.")

Space Technology Laboratories Inc. (subsidiary), 1 Space Park, Redondo Beach, California. Handles most astronautics business, including military missiles, satellite design, and prime-prime services.

Tapco Division, Cleveland 17, Ohio. Also important in aerospace work, including Dyna-Soar program management and research into advanced military space systems.

Technical Help Wanted

For the scientist or engineer with management/administrative talent, TRW offers top-notch opportunity to become a department head or the leader of a brainpower team. Its rank-and-file personnel are also hand-picked for imagination and boldness of vision into the future of astronautics. STL is among the first-rank research establishments, along with such glittering "think factories" as MIT and BTL. For the brilliant and creative-oriented grad, who might be bored with any routine space tasks, STL is his haven for mental romping into prime-prime frontiers of the unknown.

Prospects of Corporation

Geared to the Space Age in its own specialized way, TRW and its astronautics divisions are on a par with Thiokol and Aerojet-General. Though today, because of political pressure against USAF alliance with private industry, many prime-prime tasks now go to DOD-controlled Aerospace Corporation, TRW is still indispensable for its tremendous know-how in America's over-all space program. A drop in business between 1959 and 1960, caused by the above changeover of prime-prime responsibilities, was only temporary and TRW's sales graph is again rising steeply as both the USAF and NASA thankfully lay their most difficult space-system projects at this firm's doorstep. Space without TRW/STL is as unthinkable as the auto industry without Ford.

RADIO CORPORATION OF AMERICA

30 Rockefeller Plaza, New York 20, New York

Profile

Founded 1919. Besides aircraft makers, chemical firms and public utilities, RCA represents another category of big business that has thrown its hat into the space ring—the communications giants. Though its major income will continue to come from other areas —its broadcasting network, manufacture of radio and TV sets, and widely diversified consumer products and services—the share from aerospace activities is approaching 25%. As of 1962, RCA was the 26th largest private corporation in America. Total employment: 95,000.

Astronautics Feats

Transmitter for first "voice from space" (Project Score, December 18, 1958) when President Eisenhower broadcast his Christmas message via tape from an orbiting Atlas vehicle. RCA's TV system in TIROS-1 proved feasibility of weather satellites scanning earth's cloud cover daily. Echo's beacon (radio signal) allowed precise tracking of the giant balloon-satellite and gave accurate new values for the ionosphere's density. Relay-2, second active ComSat (after Telstar), is now exploring new techniques of future worldwide communications network.

Main Contracts

Prime for TIROS WeathSat series, and sub-prime (under Goddard Space Flight Center, NASA) for follow-on Nimbus series, which will lead to final operational Aeros system. Prime for Relay ComSat series to be lofted for several years. Sub-prime for Ranger lunar probe TV-scan systems (8 vehicles scheduled ahead). Subcontractor for various computer-guidance/communications systems of Mercury, Gemini and Dyna-Soar Projects. For DOD, prime for USAF's Satellite Inspector (of hostile orbiters; formerly SAINT Program), BMEWS missile (and space vehicle) detection system. Sub-prime and subcontracting roles for Bomarc, Minuteman, Hawk and Polaris missiles. Prime for USAF/NASA Project SERT (Space Electric Rocket Test) vehicles, to prove-out advanced ion-drive rocket propulsion.

Divisions

Defense Electronic Products, Front and Cooper Sts., Camden 2, New Jersey. Computers and electronic devices for military missiles.

Data Systems Division, 8500 Balboa Blvd., Van Nuys, California. Computers, data-processors, electronic warfare devices, radar equipment, much of it tailored to missile/space requirements.

Moorestown Missile & Surface Radar Division, Marne Highway, Moorestown, New Jersey. Advanced radar and range-finding systems, applicable to detection of ballistic, orbital or escape-velocity vehicles.

Aerospace Communications & Controls Division, Front &

Cooper Sts., Camden 2, New Jersey. Basic components for aerospace radar, radio, TV, detection, monitor, sensor and automation systems. Research into future orbit-rendezvous techniques, secret-code satellite messages, manned space-station AICBM maneuvers and hostile-satellite inspection.

Astro-Electronics Division, Princeton, New Jersey. Complete satellite construction (Tiros) plus space propulsion, inboard power and auto-control systems.

Major Systems Division, Marne Highway, Moorestown, New Jersey. Large-scale military missile electronics (Minuteman, Polaris, etc.), Relay ComSat construction, Dyna-Soar and space-defense sensors.

Surface Communications Division, Front and Cooper Sts., Camden 2, New Jersey. Besides earth-girdling microwave and broadcast systems, R&D into future long-range space communications.

Other major facilities include:

RCA Space Center, research laboratory for Astro-Electronics Division, opened in 1958 and greatly increased in size during 1962 to accommodate over 2000 top-rank scientific and engineering personnel who carry on advanced research in space technology.

David Sarnoff Research Center, Princeton, New Jersey. Pure and applied research, mainly in electronics and related areas of physics. Nearby, first-class laboratories of other firms and institutions have also gone up since the Space Age began, making the Princeton area one of the prime pools of "frontier brainpower" in America—or the world.

Technical Help Wanted

Research scientists in super-conductivity, pyrogenic and cryogenic phenomena, plasma physics, energy conversion, thin-film electronics, bionics, self-organizing and self-programming computers, data-processing theory, anthropo-computers (with humanlike vocal, visual, intellectual and reasoning attributes). Engineers in communications, radio, TV, microwaves, radar, with emphasis on astronautics applications. Space engineers for satellites, probes, manned stations, spaceships.

Prospects of Corporation

As one of the appointed members of the first "private space industry"—the ComSat Corporation, in joint cooperation with the government—RCA and other international carriers are pioneering in "commercial space" possibilities. Profits may not come in for 5 or 10 years, but their worldwide microwave network of orbiting "switchboard satellites" (handling radio, TV, telephone, telegraphy, fotofacsimile and data-processing transmission) will pay off handsomely from then on. Besides this, RCA will continue its healthy contracts for NASA and DOD projects extending to 1970 and beyond, with new ones not yet announced. RCA has declared that from 1963 on it will deliver a minimum of two satellites or space vehicles a month as finished hardware, plus a major astro-electronics system each week. Like its giant industrial cousins— GE, Chrysler, North American—RCA foresees space sales as someday equaling, or surpassing, earth sales.

11 : BACKUP BRIGADE

(*Statistics and background of more aerospace primes and/or sub-primes . . . Hughes . . . Grumman . . . Goodrich . . . Ford Aeronutronic (and Philco) . . . Atlantic Research . . . Avco . . . Marquardt . . . Northrop . . . Bell Telephone Labs (AT&T) . . . Raytheon . . . Minneapolis-Honeywell . . . Garrett.*)

And still, in the previous two chapters, we have not exhausted the list of primes and sub-primes in U.S. industry whose names will be forever associated with the start of the Space Age. For lack of room only—not because of lesser space status—the firms reviewed ahead must be given a more compact and less detailed rundown.

HUGHES AIRCRAFT COMPANY

Florence and Teale Sts., Culver City, California

Profile

Founded 1944. Active in missile/space field since 1959. Total personnel: 42,000. "Shangri-La" Research Laboratories (Malibu, California) accommodates 500 scientists doing advanced astronautics studies.

Milestones

Development of basic LASER (Light Amplification by Stimulated Emission of Radiation) in May, 1960, a major breakthrough in energy-wave transmission. Many important applications possible in space communications, tracking, exotic "photon" propulsion and as "powerline" to transmit solar energy from space to earth.

Main Aerospace Contracts

Falcon series of USAF air-to-air missiles; Surveyor vehicle for lunar exploration; Syncom ComSat for 24-hour orbit (prime for all).

Future Prospects

Specializes in off-trail and exotic concepts—"mobots" (mobile space robots), "plastinauts" (dummy astronauts) for space-medicine tests, "molecule manipulation" to produce new fuels such as organo-metallic compounds. Some of these will undoubtedly bring in a harvest of space contracting in the near future.

Divisions

Aerospace Group (Culver City, California), El Segundo Division (El Segundo), Field Service & Support Division (Los Angeles), Tucson Division (Arizona).

Technical Job Opportunities

Engineering scientists, Ph.D. physics researchers, electronics specialists, *all types of engineers and technicians.*

GRUMMAN AIRCRAFT ENGINEERING CORPORATION

Bethpage, Long Island, New York

Profile

Main business in aircraft. Astronautics percentage rising rapidly since 1960. Has invested heavily in space-research facilities—manned flight simulator, space environment chamber, lunar laboratory, "clean room" for miniaturization techniques.

Milestones

Developed Mercury program sea-recovery techniques.

Main Aerospace Contracts

Prime for OAO satellite (Orbiting Astronomical Observatory), to be first optical telescope in space (1964). Launch cannisters for pre-inflated Echo-2 (Big Shot) satelloon. In Apollo Program, won prime-contractor award for LEM (Lunar Excursion Module), popularly called the "Bug"—2-man vehicle that will separate from the Apollo moonship to land the first American astronauts on the moon. Also a dozen USAF and NASA feasibility studies (besides some 50 research projects of its own).

Future Prospects

With the steady decline in military aircraft sales, Grumman is after bigger space contracts in areas such as maneuverable spacecraft, manned lunar rovers, interplanetary route analysis, low-thrust nuclear-electric propulsion, unlimited restartable rocket engines.

Divisions

None.

Technical Job Opportunities

Each new hardware contract (such as the recent LEM, in 1962) requires the immediate hiring of 600 to 1000 scientists, engineers and technicians, and ultimately double that when R&D becomes production. After personnel retrenchment from loss of aircraft work, Grumman is again increasing its employment for its growing astronautics business.

B. F. GOODRICH CO.

500 S. Main St., Akron, Ohio

Profile

Like Thiokol, this firm pioneered in new synthetic and plastic solid fuels for rockets. Entered military missile field right after World War II and branched into space chemicals and materials before Sputnik. Astronautics laboratory facilities have doubled since 1960. Employees: 45,000.

Milestones

Produced first "spacesuit" (high-altitude pressure suit for Navy fliers), worn by Wiley Post in 1934, from which stemmed modern astronaut spacesuits. Introduced several new polymerized synthetic fuels that jumped missile power considerably. Propulsion for Loki made it earliest rocket in quantity for upper atmosphere investigations.

Main Aerospace Contracts

Fuel-mixes and casings for military missiles. Propellant engines for Loki, Asp and RTV rockets, both military and research. Spacesuit materials for Mercury and Gemini astronauts. Various electronic components in NASA and DOD rocketry.

Future Prospects

Goodrich is researching what will be a big field in the near future —inflatable space structures, a technique introduced by the famous Echo satellite. Plastic foams, elastic glass wools, metal-coated rubberoids and the like will eventually form re-entry drag devices (Rogallo Wings), orbital vehicles, giant space stations and housings for lunar or interplanetary camps. Goodrich fuel experts are also making breakthroughs into exotic plastic plus metallic-powder mixes that will dramatically raise the thrust ratings of solid-fuel rockets.

Divisions

Goodrich Aerospace & Defense Products (500 S. Main St., Akron, Ohio), Goodrich Chemical Co. (3135 Euclid Ave., Cleveland 15, Ohio), Goodrich High Voltage Astronautics (Box 98, Burlington, Massachusetts).

Technical Job Opportunities

Primarily chemical engineers and scientists, but also all branches relating to rocket propulsion, astro-physics and bio-astronautics.

FORD MOTOR COMPANY (Aerospace Only)

The American Road, Dearborn, Michigan

Profile

This well-known auto firm followed Chrysler's lead and in 1956 plunged headlong into astronautics, when its Aeronutronic Division was organized. Acquired Philco as a subsidiary in 1961, greatly expanding its electronics aerospace capacities. Aims for important manned spaceflight contracts. Aeronutronic personnel: 3000 (50% scientists and engineers).

Milestones

In 1956, Far-side rocket, carried up 100,000 feet by balloon, ignited and broke altitude records by going 5000 miles. Philco feats were the first transistors in a space vehicle (Vanguard I, March, 1958), and the first active (powered) radio-relay communications satellite (Courier, October, 1960).

Main Aerospace Contracts

Aeronutronic Div.: Prime for USAF's Blue Scout Jr. launch rocket and for Army's Shillelagh missile. Sub-prime for Lunar Ranger's "talking ball" (soft-landing capsule); re-entry of MMRBM (Mobile Medium Range Ballistic Missile); ICBM decoys. Philco: Sub-prime in Mercury astronaut tracking system; Sidewinder missile guidance; Redeye missile target-seeking (infra-red) system.

Future Prospects

Aeronutronic also has NASA study-contract to devise life-detection sensors for planet probes (to land on Mars and Venus in 1964-65). For USAF, various classified projects in space defense. Among company-financed astro-researches is BIAX, a memory-bank unit for the super-computers that will be vital for astrogation of spaceships and other mind-staggering astronautics feats to come. Corporation business in aerospace is expected to multiply several times by 1970.

Divisions and Subsidiaries (Aerospace Only)

Aeronutronic Division (Ford Road, Newport Beach, California), Communications & Electronics Div. (4700 Wissahickon Ave., Philadelphia 44, Pennsylvania), Scientific Laboratory Facility (Union Meeting Rd. & Jolly Rd., Blue Bell, Pennsylvania), Philco Corporation, subsidiary ("C" and Tioga Sts., Philadelphia 34, Pennsylvania).

Technical Job Opportunities

Engineers for all types of aerospace hardware—missiles, spacecraft, satellites, probes, space stations. Engineering and research scientists in exotic electronics. Technicians in the total aerospace arena.

ATLANTIC RESEARCH CORPORATION

Shirley H'way & Edsall Rd., Alexandria, Virginia

Profile

This is the smallest of the aerospace firms we've dealt with so far —and one of the fastest growing. Founded in 1949, its sales reached $1 million in 1953, $2.5 million in 1957, and $8.3 million in 1959. Its highly competent staff of about 4000 is 50% professional, most with advanced degrees rather than the B.A. or B.S. Geared solely to the age of rocketry (like its big brother, Aerojet-General), this firm has become perhaps the top-rated sub-contractor in the field, sought by all government and industry primes for its class-A hardware and scintillating research. It is second only to Marquardt in the research rocket field, having developed ten of these by itself. The "little giant" of astronautics, Atlantic Research is truly a model for all other aerospace firms, small or large.

Milestones

Indirectly, ARC shared in the phenomenal success of the "poor man's rocket"—USAF's Blue Scout Jr. solid-fuel 4-stage space booster. Also developed the important Trailblazer-2, which uses four stages to go up into space and two stages to "power dive" back into the atmosphere at high velocity, for re-entry research at interplanetary return speeds (25,000 mph and over). One of the most remarkable reliability records was established in 1961 when, at Eglin AFB, Arcas-Robin high-altitude rockets were shot off each hour—24 in 24 hours—without a single failure, and all reaching their designed height of over 240,000 feet (average for all: 244,000 feet).

Main Aerospace Contracts

Prime for sounding rockets Iris, Nike-Archer, Archer, Boosted Arcas, Arcas, Metroc (METeorological ROCket), Boosted Metroc, Arcon, Trailblazer-1, Trailblazer-2. Metroc is an all-plastic rocket, casing included. Sub-prime for Blue Scout Jr. Subcontracts for various military missile components (Redeye, Terrier, Polaris, Skybolt, Titan, Nike-Zeus, Minuteman) and for Mercury capsule auxiliary rocket systems (tower-jettison and posigrades separating capsule from booster). Several dozen research and feasibility studies, from the USAF and NASA, are also conducted yearly.

Future Prospects

If its spectacular rise continues (and no slightest evidence points the other way) ARC will within this decade pass from small to big business and in time may become one of the aerospace giants. Two of its unique new concepts—gel-solid ("paste") rocket fuel that is automatically formed and loaded, and spherical rocket-engine units—are under top-priority R&D with USAF funding. ARC's extensive production of meteorological and specialty research rockets is likely to be the steadiest of steady business into the future (the 1000th Arcas was fired in early 1963). As with STL's "prime-prime" services, and McDonnell's manned vehicles, ARC has cut out its own private niche as an indispensable cog without which the nation's whole space machine would be hampered. Its laboratories and plants are strategically situated around the country (New England, New York, Pennsylvania, California, Florida) and run with superb central integration. If there is any such thing as a "pure product" of the Space Age in the astronautics industry, it is this small but potent firm.

Divisions

Several smaller acquired businesses, but all under the centralized management of ARC's home office.

Technical Job Opportunities

ARC has a very excellent approach to recruiting its rapidly expanding staff (about 35% increase per year). Inquiry at their Professional Personnel Recruitment Office will bring you full brochures of their projects, plus detailed listings of their current manpower needs, specifically naming each type of engineer, scientist or technician. A generous scholarship and on-the-job training program is open for new company members. They list their main fields of science technology as R&D of high-energy propellants, materials technology, nuclear energy research, slurry (semi-fluid) fuels, combustion and heat-transfer, and astrodynamics. Their products, besides rocket systems, include transducer equipment, calibration instruments, data conversion units, antennas, telemetry components and signal generators. A strong research team has recently been committed under USAF contracting to classified work in electronic warfare methodology (which goes beyond present-day missiles into exotic orbiting defense systems and space-based manned maneuverable craft).

A final word: It is not easy to join ARC's top-grade staff. Their standards are on a par with those of NASA itself. But those who make the grade are assured of maximum creativity outlets, smooth working relationships, ultra-modern laboratory equipment and as high a scale of pay as may be found anywhere in the astronautics industry.

AVCO CORPORATION

750 Third Ave., New York 17, New York

Profile

This firm was one of the first industrial research organizations in America, starting at the turn of the century with farm machinery. Today its diversified R&D activities have spread into electronics, missiles, space systems, aircraft, radar, ordnance, gas turbines and radio and TV equipment. Of its gross today, 80% is derived from the aerospace field, more of it from R&D than from hardware. "Frontier research" will continue to be Avco's main "product."

Milestones

(None "spectacular" but all highly important to the advance of missiles and space technology): First ablative ("peel-away") nose-cone re-entry success (January, 1961; Minuteman). First spray-on ablative heat-shield coating and first quartz-honeycomb structures for nose-cones. Astronaut-style centrifuge test-chamber (for USN before 1957). First pilot MHD (*magnetohydrodynamics*, now shortened to *hydromagnetics*) power generator via super-heated plasma (thin ionized gas) in 1961.

Main Aerospace Contracts

Sub-prime for Atlas, Titan and Minuteman nose-cones (alternately with GE on certain rocket series); for Samos re-entry system; for Apollo spacecraft heat shield; for military missiles Alfa and SD-2 (latter a unique drone-missile, for unmanned battle-field reconnaisance). Subcontracts for Nike-Hercules, Talos, Bullpup. Major R&D and study contracts for NASA and DOD in advanced MHD power-systems, re-entry physics, monitoring re-entry ICBM warheads, metallo-plastic honeycomb structures for space-craft, destruct of space nuclear reactors, many more.

Future Prospects

Avco's specialized research services cannot be duplicated elsewhere, hence have become a vital link of the space chain. It is now active in more than 30 missile and space projects. Important advanced space concepts pursued in its labs—infra-red space communications, MHD propulsion, interplanetary guidance and terminal command-control—are receiving increased appropriations each year for the Apollo and post-1970 space travel period. Avco is one of the two giants (with GE's MSD) of advanced nose-cone and re-entry technology, which will become a top-priority mission when manned spaceflights become more numerous and piloted ships seek return through earth's "air barrier" on a monthly or weekly basis. Via its steady breakthroughs, Avco will continue to be in the space frontiers with the best of them.

Divisions

Avco-Everett Research Laboratory (2385 Revere Beach P'way, Everett 45, Massachusetts), Avco Electronics & Ordnance Division (Cincinnati 15, Ohio), Lycoming Division (550 S. Main St., Stratford, Connecticut), Nashville Division (Box 210, Nashville 1, Tennessee), Research & Advanced Development Division (201

Lowell St., Wilmington, Delaware), Crosley Broadcasting Corp. Division (Cincinnati, Ohio), Avco New Idea Division (Coldwater, Ohio; Sandwich, Illinois; Ft. Dodge, Iowa).

Technical Job Opportunities

Particularly good for research scientists in space technology, MHD phenomena, re-entry dynamics, exotic propulsion (Avco-Everett Lab). Interplanetary space vehicles, future Space Age materials, all-octave space signals, bionics (R&D Development Lab). Advanced space electronics in classified USAF projects, deep-space control and tracking techniques, specialized space-defense satellites and manned orbiters (Avco Electronics & Ordnance Labs). Much of this "applied" research is, in reality, "pure" research of the most fundamental kind. If you want to take "untrodden paths," join Avco's breakthrough brigade.

THE MARQUARDT CORPORATION

Corporate Offices, 16555 Saitcoy St., Van Nuys, California

Profile

Founded 1944 as an aircraft-engine maker, it turned to aerospace work in 1946, developing supersonic jet engines for military missiles (Hound Dog, Bomarc), and becoming the peerless high-velocity turbojet and ramjet specialist in America. Ending of the Bomarc Program (replaced by solid-fuel Mace) in 1961 brought a drop in sales, but new projects are now bringing the gross up again. By acquiring the Cooper Development Co. as its division (1960), Marquardt became tops in the research rocket field (total of 12). Main business is still in military jet engines (USN Phantom, various target-drones, trainers).

Milestones

America's first ramjet engine drove the X-7 unmanned plane (testbed for Bomarc) beyond Mach 3 (2000-plus mph) and to highest altitude (100,000 feet) for air-breathing missile. By 1960, the X-7 with its recoverable system (and called "Old Boomerang") had racked up 13 flight missions. Through Cooper, the corporation holds a dozen high-altitude rocket records. Recently devised first "hyperjet" engine combining both rocket and jet power.

Main Aerospace Contracts

Prime for Cooper rockets—Asp I and IV, Aspan 150 and 300, Boa, King Cobra, Queen Cobra, Python, Roksonde 100 and 200, Starseeker, Ascamp. Sub-prime, USAF's Pluto nuclear-ramjet (for SLAM low-flying, continuous-powered ICBM), for Army's Red-head-Roadrunner missile (supersonic target missile), for USAF's ALBM Hound Dog. Subcontract for Apollo spacecraft's on-and-off propulsion unit and for lunar LEM's attitude controls. Also innumerable jet/ramjet engine components and instrumentation.

Future Prospects

Ramjet technology will become a major item when recoverable booster systems (for Titan-3, Saturn, Nova) are incorporated from 1964 on. Quick-reaction jet-controls of Marquardt will find many uses in orbit-maneuver vehicles, mooncraft landings and interplanetary mid-course guidance of spaceships. The DOD/AEC SLAM missile may become top-priority if NASA's Project Rover nuclear rocket proves workable (around 1965). Recent R&D contract for propulsion of USAF Aerospace Plane, step after Dyna-Soar, with ground take-off and landing capability after extended space tours around earth. Stable of a dozen weather and atmosphere research rockets will be ordered in rising numbers as hitherto, by USAF, NASA and U. S. Weather Bureau.

Divisions

Power Systems Division (same address as parent corp.), Ogden Facility (1000 W. 33rd St., Ogden, Utah), Pomona Division (2709 N. Garey, Pomona, California), Special Projects Laboratory (Warehouse 4A, Congressional Airport, Box 723, Rockville, Maryland),

ASTRO Division (Air/Space Travel Research Organization, home address), Cooper Development Div. (Monrovia and Van Nuys, California).

Technical Job Opportunities

Propulsion and spaceflight engineers are the primary help wanted. But other avenues are open for nuclear-electric propulsion (plasma diode and "resistojet" systems are Marquardt specialities), pyrogenic metallurgy, advanced astro-electronics, zero-g fuel phenomena, nuclear dynamics, high-speed cryogenic gas-fluid separation and other esoteric technology related to the future powerplants of our wide-ranging manned and unmanned spacecraft.

NORTHROP CORPORATION

Box 1525, 9744 Wilshire Blvd, Beverly Hills, California

Profile

Founded in 1939 as an aircraft manufacturer. With Snark contract (jet-powered ICBM) in 1950, began switching heavily toward aerospace field, which comprises 75% of gross today. Corporation has extraordinary diversification through its divisions so that it has fingers in almost every space pie (over 75 different areas of missiles and space vehicles). Its sales have shown a consistent, often steep rise, both before and after the Space Age began in 1957. Total employees: over 22,000 (including over 4000 science technologists).

Milestones

Basic astrogation (space guidance) system of Mariner-2 of Venus flyby fame (December, 1962). First ICBM vehicle (jet-powered Snark) prior to rocket-powered Atlas and Titan. Guidance device for Ranger-4, which struck moon (April, 1962).

Main Aerospace Contracts

Prime for rocket-powered target drone of all military services; for Army's surveillance drone. Sub-prime for guidance of Skybolt and Hawk missiles; for Samos ReconSat. Subcontracts for parts within Polaris, Titan, Lacrosse and a dozen other military missiles; for recovery systems of Mercury, Gemini and Apollo Programs; for X-15's "Q-ball" (re-entry analyzer); for all Ranger/Mariner/Voyager interplanetary probes. To become prime for all lunar Ranger vehicles in 1965, according to NASA.

Future Prospects

By constantly forging into unexplored technological frontiers, Northrop is assuring itself of always-increasing space business—a Lunar Center Finder to accurately guide the first moon-landing astronauts, novel applications of the LASER to tracking systems, orbit-rendezvous short-cut systems and a host of others. Its widespread military-missile coverage, satellite and probe guidance, and manned spaceflight contracts are bound to multiply Northrop's gross year by year.

Divisions

Northrop Space Laboratories (1111 E. Broadway, Hawthorne, California), Norair Division (1001 E. Broadway, Hawthorne, California), Nortronics Division (Research Park, Palos Verdes Estates, California), Ventura Division (8000 Woodley Ave., Van Nuys, California).

Technical Job Opportunities

New recruits have the special advantage of free academics and job-training at the famed Northrop Institute of Technology (which graduates many more technologists for all other aerospace firms as well). Hardly any other aerospace firm boasts a wider variety of scientists, engineers and technicians than Northrop and its divisions. Research fields include variable geometry, functional space vehicles design, extraterrestrial excursion vehicles (surface exploration), interplanetary outpost shelters, astrogation systems throughout solar system, earth-orbit rendezvous, lunar-orbit/Mars-orbit/Venus-orbit systems, electrostatic space propulsion, zero-g bio study, high pressure electronics, anti-meteoroid techniques. Obviously, from this small portion of their total range of investigations, Northrop is leaving no space stone unturned.

BELL TELEPHONE LABORATORIES INC.

463 West St., New York 14, New York

Profile

A subsidiary of AT&T, the communications giant that, like RCA, is part of the Space ComSat Corporation in process of establishing a world microwave network. BTL handles almost all the aerospace R&D contracting of the parent firm, and along with MIT, Johns Hopkins, JPL and STL shares world honors for its top-calibre research. In fact, many experts rate BTL as the world's greatest, for this partial list of its phenomenal breakthroughs—the transistor, MASER (parent of LASER), solar cell power conversion, radio-telescope—all of which have profoundly stimulated space research and technology. The 5000-man brainpool of BTL is, in effect, behind almost every piece of space hardware.

Milestones

First solar-cell satellite radio transmissions and first space application of transistors (Vanguard I, March, 1958). First deep-space tracking antenna (Goldstone, California) based on radio-telescope. First long-distance telemetry pickup via super-sensitive MASER receiver (Pioneer-IV, March, 1959) from trans-lunar range of 400,-000 miles. First space-relay of signals from passive satellite (Echo-1, August, 1960). First moon-bounce signals (1960). First LASER "spotlight" on the moon (1962). Telstar alone racked up a long list of Space Age firsts—transatlantic radio, telephone calls, television (both black-and-white and color), telegraphy, ultra-speed computer-data transmissions, newspaper and photo reproduction, teletype, intercontinental network-TV hookup (America and 17 European nations).

Main Aerospace Contracts

As with RCA, AT&T is bound up with the government in the international ComSat project, for which BTL is prime for producing experimental Telstar satellites (but at its own expense, plus paying for launching costs). Sub-prime for guidance of Thor-Delta and Thor-Ablestar launch vehicles, which both have a remarkable reliability record in orbiting satellites (13 in a row for Delta without an abort). Inertial guidance for all Titans, both as missiles and space boosters (to include Titan-2 and Titan-3). Target-seeking techniques of Nike series of rockets (along with Douglas). Radio-bounce equipment for forthcoming Echo-2 (Big Shot) balloon satellite. Also, under a special contract, BTL has "loaned" some 200 of its top scientists and astro-engineers to NASA as an over-all "space program management" body to fully integrate all phases of our national astronautics effort into a smooth-running entity.

Future Prospects

As before, BTL will continue to "make its own future" simply by peerless pioneering into science fundamentals and unique aerospace technology. Telstar's overwhelming success has already assured BTL its firm place in earth-girdling space communications in this century, and the next, no doubt. Their research repertory may lead to great new breakthroughs in the following fields—space-to-earth LASER-beam (wireless) transmission of solar power; some form of anti-gravity "shield"; LASER "ray-gun" space weapons; receiving messages from other inhabited worlds in outer galactic space; detecting life on Mars; atomic-clock satellite experiment to test Relativity Theory; creating anti-matter and unlocking anti-matter plus norm-matter (mutual annihilation) power (1000 times greater than nuclear energy). Even more stunning revelations (who could foresee even the transistor?) will more than likely come out of BTL's doors and shake the world.

Divisions (of AT&T)

Besides its non-aerospace facilities, the parent corporation has one other subsidiary holding aerospace contracts—Western Electric Co. Inc. (222 Broadway, New York 38, New York), which primarily produces the hardware first researched at BTL. Western is the prime for Nike-Ajax, Nike-Hercules and Nike-Zeus as complete missiles; sub-prime in guidance or electronics devices for Titan and the Nike series. BTL has its own divisional branches scattered through the Northeast—at Holmdel, New Jersey (tracking/receiving station); Murray Hill, New Jersey (main research labs); Andover, Maine (Telstar space-signal station); and others.

Technical Job Opportunities

Besides scientists and engineers of "genius" grade, the less-creative though not less important types of applications, production and design technologists are welcomed as BTL's staff steadily grows through the years. For the "master-mind" honor grads with a powerhouse of new science ideas packed between their ears, BTL is the place to unload them on the gasping world. As one example: Dr. John R. Pierce was primarily responsible for: the Echo radio-relay concept, the traveling-wave tube for low-power space transmission from orbital heights, the fantastically ingenious microwave equipment of Telstar, and the horn-antenna concept for space-signal receptions (not to mention a half dozen books on basic message and signal-code theory that are the "gospel" of space communications as known today). Such Space Age Edisons and Einsteins are carefully nurtured into full flower at BTL.

At Western Electric's plants, design and production engineers are on the want-lists, for military missiles and space hardware. Technicians in all categories are in demand, as are applied scientists and technologists specializing in rocketry.

Spring St., Lexington 73, Massachusetts

Profile

Founded 1925 and one of the first firms in the then-new electronics field. Entered missiles/space competition in 1956 and grew phenomenally, now being among first 100 giants of U.S. industry. Pioneered in radar, miniaturization, computers and guidance. Total employees: 45,000.

Milestones

Initiated concept of "packaged electronics" (micro-miniaturized modules), which greatly aided guidance of missiles and space rockets. Its guide-controls for Army's HAWK (Homing All the Way Killer) made it top operational missile (used by entire NATO forces). Breakthrough in inertial guidance reduced Polaris unit to one-tenth former size, leading to longer-range A-2 and A-3 models.

Main Aerospace Contracts

Guidance systems of HAWK, Sparrow-3, Tartar and Polaris. Compact computers for guidance of Apollo manned craft in lunar program. Also electronics components in many other rockets and spacecraft.

Future Prospects

As a leader of the "electronics explosion" Raytheon holds a vital role in our present and future space effort. Their newly organized "space braintrust" is researching the fantastic field of "molectronics," wherein "giant" transistors will be partly replaced by microscopic molecules themselves (of semi-conductors) and usher in the era of "vest-pocket computers" and "shoebox powerplants." Raytheon scientists (and others) predict that the greatest electronic marvels are still to come and that today we are on the verge of a staggering breakthrough in harnessing subatomic powers.

Divisions

Laboratories and plants scattered through the Greater Boston area represent miniature divisions tightly integrated with the parent corporation—Research Division, Machlett Laboratories, Industrial Components, Equipment Division, Semiconductor Division, Microwave and Power Tube Division, Electronic Services and several non-aerospace plants. Missile & Space Division (Highway 128, north Boston's Electronics Row) is the newest and fastest growing.

Technical Job Opportunities

Specialists and engineers in all areas of the enormously splintered electronics field: solid-state physics, miniaturization, printed circuitry techniques, MASER and LASER research, infra-red and exotic microwave generation, ultrasonics, high-heat operable electronic devices and cryogenic electromagnetic phenomena. Electrical engineers and research physicists can practically create their own new field of work classification by striking in new directions in the vast electronics frontier.

THE GARRETT CORPORATION

9851 Sepulveda Blvd, Los Angeles 9, California

Profile

Founded 1936. Dominates life-support field (pressurized or sealed environments) for aircraft, missiles and space vehicles. Has remarkable record of installing life-support units in every high-altitude vehicle America ever had (total of 175 aircraft). Drop in sales from 1960 to 1961 represented cut-back in military planes, for which Garrett supplied super-chargers and other devices. But space contracts are coming in strong now. Employees: 11,500.

Milestones

First pressurized cockpit (B-29 bomber of World War II) in aircraft, and first life-support system in spacecraft (Mercury capsules of astronauts). By a 10-to-1 reduction in size/weight factors for life-support unit, made possible the 24-hour flight of Cooper's Mercury craft in 1963.

Main Aerospace Contracts

Prime for life-support systems in Mercury flights, and for all future Gemini and Apollo craft up to moon landings. Sub-prime for environment system of X-15 and Dyna-Soar. Pressurization systems of Discoverer satellites; of Pershing, Sergeant, Subroc and Titan missiles. Components in two dozen other missiles.

Future Prospects

Garrett will most likely continue to dominate space life-support techniques when orbiting laboratories, space stations and lunar camps are established before and after 1970. Firm is already researching cryogenic gas-separation for USAF's Space Plane concept (beyond Dyna-Soar) and for Project BOSS (Bioastronautical Orbital Space System), a "test-bed" for military astronauts of the future. Also is important subcontractor in SPUR Program (Space Power Unit Reactor) of USAF, which may become a big project for onboard power-systems of manned space vehicles of all kinds.

Divisions (missiles & space only, excluding aircraft):

AiResearch Division (industrial plant: 9225 Aviation Blvd, Los Angeles 45, California; Mfg. plant: 9851-9951 Sepulveda Blvd, Los Angeles 45, California); AiResearch Mfg. Co. of Arizona (402 S. 36th St., Phoenix, Arizona).

Technical Job Opportunities

Garrett predicts doubling its labor force in five years. Most additional personnel will be in the space area, particularly bio-astronautics, cryogenic gases, life-support servo-mechanisms and human engineering techniques for astronauts. However, Garrett is winning research grants, feasibility studies and R&D contracts in widespread fields—rocket propulsion, nuclear/electric drives, spacecraft design, re-entry cooling systems, pressurization of inflatable space stations, and many more.

MINNEAPOLIS-HONEYWELL REGULATOR COMPANY

2747 4th Ave., South Minneapolis 8, Minnesota

Profile

Founded 1885. Business turned to parts for aircraft in 1941, to missiles and space in 1956. MHR is the "guidance king" because of its gyro and inertial stabilizer systems, which are in 75% of all space vehicles launched by America to date (including 50 Discoverers alone; also Midas, Explorer, Pioneer, Transit, Courier, Echo and Tiros). Total employment: 60,000.

Milestones

Basic originator of the first inertial guidance system* for Vanguard Project (1956-1960). First frictionless gyro utilizing whirling ball (by electrostatic forces) in a vacuum casing (1960), now used in many launch rockets and military missiles. First "clean room"—ultra air-conditioned and dustless for assembling precision rocket devices—in 1951. First Adaptive Flight Controls (automatic jet-thrust system) for X-15 in 1959. First electronic and first supersonic auto-pilot, for X-1A flight that broke the sound barrier in 1947. To date, MHR has produced more than a half million gyro and accelerometer components used in U.S. missiles and space vehicles. As with Northrop and Avco, hardly a single piece of Space Age hardware is launched that does not contain some kind of MHR device or system.

* Based on inertia and independent of the earth's gravitation.

Main Aerospace Contracts

Prime for ASROC (Anti Submarine ROCket; i.e. rocket-driven torpedo) of USN. Sub-prime for guidance (whole system and/or gyro and/or inertial stabilizer platform and/or accelerometer) of Mercury capsules, Gemini craft, X-15 rocketplane, Dyna-Soar spacecraft, Agena upper-stage satellite unit, NASA Scout and USAF Blue Scout space launchers, Thor-Agena and Atlas-Agena launch vehicles, Sergeant missile. Subcontract devices (such as horizon-sensors) in over 100 other U.S. missile/satellite/space projects.

Future Prospects

MHR is not likely to be dislodged from first place in aerospace guidance contracting, both for the USAF and NASA. The Mercury/Gemini gyro-jobs undoubtedly mean some big part in the Apollo spacecraft astrogation systems, and for interplanetary ships beyond. Sideline techniques of MHR in servo-mechanisms, computer systems and miniaturized electronics will also reap their share of astronautics business into an indefinite future. Space is here to stay and so is MHR.

Divisions

Aero Division (2600 Ridgway Road, Minneapolis 13, Minnesota); also plants in Florida, Los Angeles, and Canada; Boston Division (1400 Soldiers Field Rd., Boston 35, Massachusetts), Ordnance Division (600 2nd St. N., Hopkins, Minnesota); also plants in California and Washington State; Brown Instruments Division (Wayne & Windrim Ave., Philadelphia 44, Pennsylvania); M-H Heiland Division (Box 8776, 4800 E. Dry Creek Rd., Denver 10, Colorado); Micro-Switch (11 W. Spring St., Freeport, Illinois); Precision Meter (Grenier Field, Manchester, New Hampshire); Rubicon Instruments (Ridge Ave. & 35th St., Philadelphia 32, Pennsylvania); Semiconductor Products (1015 S. 6th St., Minneapolis 4, Minnesota); Special Systems (Queen & Bailey Sts., Pottstown, Pennsylvania).

Technical Job Opportunities

Engineers and technicians in areas of basic mechanics, gyro dynamics, torque fields, physics of frictional forces, and centrifugal dynamics. Also, technologists in all related electronics fields. MHR's research roster is constantly being augmented in order to staff new "laboratories of tomorrow," the latest being their self-styled "Technological Wonderland" for advanced aerospace studies. Wide choice of plants across the country makes MHR an ideal employer for many new recruits who desire one location over another.

The above (in the last three chapters) represent less than half of the firms who have won important roles in America's total missile/aerospace/astronautics industry. Appendix IV lists more business firms in thumbnail brevity—which has no slightest connection with how significant they are in the Space Age industrial picture. Many of them count more of their gross dollars from aerospace work than the group previously reviewed. Some have a superior bag of overflowing contracts. Others perhaps have a greater space potential.

12 : FIRMS WITH A FUTURE

(Survey of 200,000 small businesses in the aerospace field . . . What they are, what they make or do . . . Advantages of small-firm job . . . Disadvantages . . . Wide variety of science technology businesses . . . Colorful Space Age names . . . The electronics explosion and its great opportunities . . . Phenomenal growth of small space firms . . . Spectacular sales and profits . . . Domination of small-firm labor force by technical help.)

We have hardly begun to discuss the long list of firms engaged in aerospace industry. Only the comparative giants have been dealt with, leaving some 200,000 smaller sub-sub-contractors and supply firms. It is quite likely that new ones are climbing on the astronautics bandwagon faster than a typewriter can list them.

Full lists, if anyone is interested, are contained in annual or semi-annual "buyer's guide" editions of trade magazines, filling some 250 pages in fine print. However, though we cannot list them all, they should not be ignored as a group. Your best job may not necessarily lie with any of the bigger firms, but with a small business (under 500 employees in the usual definition). Good things often come in small corporate packages:

—More personal attention to you and your abilities.
—More chance to rise to the top.
—More specific duties so that you feel you are doing your ut-

most. (For any person with a lively brain, nothing is worse than enforced "idling".)

—Compact technical teams in which your identity is not "lost." (A serious personnel problem among giant firms is keeping together the huge armies of engineers required for prime contracts.)

As in everything in life, the disadvantages are there too. Subject to the well-known attrition that afflicts all small businesses, which go bankrupt each year by thousands, small firms offer less security. Often, in their early years, those that survive cannot match nationwide salaries, and one must go on future promises and hopes. One particular bane of small firms is the tendency to become "family dominated," so that a second cousin or in-law is appointed to management just when you expect and deserve it.

However, there is one small-firm factor that cannot be matched by any giant corporation. Lockheed, North American, GE, Douglas—none of them will ever double or triple their business in one year, and for years to follow. This can only happen—and *does* at times—with the small firm that comes up with some special product or ingenious bit of technology that commands the highest bidding among the prime contractors when sub-contracts are being handed out.

This writer, as editor of *Space World* (a non-technical monthly publication popularizing astronautical science and our space program), has for five years seen releases and financial statements come in from small "unknowns" who suddenly become very well known as their business skyrockets. Whether through business acumen, wise build-up of first-class technological talent, or sheer luck, all small firms connected in any way with the Space Age— even if they make only one tiny tunnel diode in quantity—have the chance to be swept up by the astronautics tidal wave. If you are with them, you also ride the crest to golden shores.

These supply and specialty firms spread through all areas of industry—chemicals, electronics, metallurgicals, gadgetry, tools and on down the line. They also deal in services, such as working out esoteric engineering techniques, compiling statistics or acting as scientific consultants. The trade magazines divide the great swarm into broad groups, usually Aircraft, Space Vehicles and Missiles, Avionics and Support Equipment and Services—with sub-headings such as Antenna Systems, Automation, Gyros, Combustion, Cryogenics, Guidance, Life Support, Nose Cones, Space Suits.

The firms that follow are a purely random sampling of those on which representative data are available. The categories they occupy are my own, to lend clarity. Their names have broken all tradition of "dignity" (i.e., colorlessness) and are exuberantly fashioned straight from space and science terminology. The drab name is as dead as a dodo in modern Space Age industry. They either spell out precisely what they produce or are coined out of intriguing technical terms:

Firm	*Product*
Magnesium Aerospace Products Bay City, Mich.	Magnesium and alloys for missiles and spacecraft.
Electrospace Corp. Glen Cove, N.Y.	Electronics communications systems.
Ortronix Inc. Orlando, Fla.	Electronics in assorted fields.
Electroid Corp. Union, N.J.	Solenoids and small electronics components.
Dynalectron Corp. Washington, D.C.	Telemetry and other space-vehicle electrical instruments.
Avien Inc. Woodside, N.Y.	Electronics parts in aircraft and spacecraft.
Astro-Science Corp. Culver City, Calif.	Satellite tape-recording and transmission equipment.
Resitron Laboratories Inc. Santa Monica, Calif.	High-voltage and high-vacuum devices.
Space Age Materials Corp. Woodside, N.Y.	Pyrogenic nose-cone products and high-temperature techniques.
Semi-Elements Inc. Saxonburg, Pa.	MASER-type metals (germanium, tantalum, neodymium, terbium, etc.).
Washington Scientific Industries Inc. Minneapolis, Minn.	Electro-mechanical systems for missiles and space rockets.

| General Time Corp. | Timing devices and data-processing systems. |
| Ether Ltd. | Temperature controls and automation systems. |

ELECTRONICS TIDAL WAVE

Under the spur of the Space Age, rocket technology is booming, missile engineering is mushrooming and bio-astronautics is leap-frogging ahead. But leaving them all far behind in the past decade is the *electronics explosion,* the truly astounding phenomenon of today. Each new gadget or circuit is constantly being miniaturized, micro-miniaturized and sub-micro-miniaturized 10 or 50 times smaller than before. At the same time, performance is not merely doubled but quadrupled twice over, or takes a quantum jump of a hundredfold.

What were originally wisecracks in electronics labs have turned into serious work slogans: "If you can see it, it's too big" . . . "It's off the drawing board, where's the Obsolete stamp?" . . . "Can you imagine, not a breakthrough for a whole week?"

Electronics is modern miracle-working. A more realistic definition narrows it down to that branch of physics dealing with applied electrical hardware for communications, power supply, radiation sensors, telemetric systems and the like. It packs radio transmitters into thimbles and computers into shoeboxes. It is an ever-shrinking Lilliput Land that has gone from former thumb-size radio tubes to pea-sized transistors to pinhead diodes to microscopic dielectrics, and is now exploring the invisible tininess of molecule "gadgetry." It has branched out into a many-splendored maze of opposites—into vacuum-thin plasma physics and densely packed solid-state physics, into jawbreaking magnetohydrodynamics and mind-breaking thermionics.

And instead of reaching any foreseeable limitation to its subatomic wonders, it seems to have broken through into a yawning electromagnetic universe more staggeringly gigantic than the astronomical universe our spaceships will explore.

The birth rate of electronics firms is higher than the population explosion of humans. Perhaps their death rate is too. But the survivors grow mightily. The majority are space-spawned and less than five years old. They usually start with a couple of ambitious young electronics experts quitting their jobs and setting up shop in an old garage with a truckload of equipment and a piggybank load of capital. Debts pile up faster than diodes at first. But with luck, and some electronics legerdemain, they break the ionic barrier and roar to Mach $2 million and beyond.

Tiny and unknown not long ago, such firms as Radiatronics, American Electronics, Straza Industries, Telex Inc., Bell Electronic Corporation, Electronic Communications, Astrex, Babcock Electronics and California Magnetic Control Corporation are already as solidly established as venerable establishments a half-century old.

Sales and profits can be as spectacular as the industry's Promethean breakthroughs:

—California Computer Products Inc. (Downey, Calif.) increased its 1962 earnings 3.8 times over 1961, while sales, somewhat "laggingly," only went up 2.4 times.

—Instrument Systems Corporation (New York City) shot from a 1961 profit perigee of $28,000 to an apogee of $455,000 in 1962.

—Electronic Specialty Co. (Los Angeles, Calif.) parlayed its $13,000 earnings of 1961 like a thoroughbred into a giddy $887,000 for 1962.

—Zero Manufacturing Co. of Burbank, Calif. (mentioned previously) went from a 1961 take of $32,000 to a lofty 1962 bracket of $450,000, and raised earnings on shares from 4¢ to 49¢—a 1200% increase that must have put shareholders into a joyful orbit.

—Astro-Science Corporation (Culver City, Calif.) bestowed an incredible 200-for-1 stock split on its dumfounded backers— a Paper Strike outglittering the California Gold Rush of a century ago.

Similar space bonanzas occur among non-electronics small businesses engaged in supplying astronautics goods or services to the giants of primeland. Firms more than doubling—and often redoubling—their 1961 profits in 1962 include:

—Washington Scientific Industries Inc. (Minneapolis, Minn.), specializing in servomechanisms.

—PneumoDynamics Corporation (Cleveland, Ohio), producing pneumatic and hydraulic control valves for space vehicles.

—Consolidated American Services Inc. (Hawthorne, Calif.), metal-processing experts.

—Superior Mfg. & Instrument Corporation (Long Island City, N.Y.), turning out servo-system devices for spacecraft.

—Lithium Corporation of America Inc. (Bessemer City, N.C.), dealing in such semi-rare wonder metals of the Space Age as lithium and beryllium.

A vast swarm of small firms dispensing intangible services with minimal hardware are also prospering:

—Geophysics Corporation of America (Bedford, Mass.) has a NASA contract to study the weather—on Mars and Venus.

—Raymond Engineering Laboratory Inc. (Middletown, Conn.) is whipping up missile computer-programs on order.

—Rosemont Engineering Co. (Minneapolis, Minn.) is developing new thermo-sensors for space use.

—National Research Corporation (Cambridge, Mass.) has a Space Vacuum Laboratory that tests rocket paraphernalia at simulated heights of 500 miles.

—Reynolds Rocket Systems Inc. (La Puente, Calif.) researches any way-far-out feasibility study, such as land-on-a-dime maneuvers suitable for visits to the moon and planets.

—Applied Electronics Corporation (Metchuchen, N.J.), which unabashedly subtitles itself a "David Among Aerospace Goliaths," proved it by beating out its big rivals in designing the 90-channel telemetry system in orbital Mercury flights, starting with John Glenn's.

—MITE Corporation (Miniature Industrial Technical Equipment, New Haven, Conn.) has "thought big in a small way" to reap mighty profits by minting mites of technology.

Modest organizations also earn magnified rewards by developing brand-new devices that instantly prove valuable and create strong demand:

—Power Components Inc. (Scottdale, Pa.) rang the bell with an epoxy diode that outperformed silicon and other semi-conductors in certain electromagnetic octaves.

—Hexcel Products Inc. (Berkeley, Calif.) conjured up an aluminum honeycomb which "telescopes" out from a 2-inch pack to a 6-foot strip, ideal as a foundation filled with foaming-plastic for forming the walls of future inflatable space stations.

—Hycon Mfg. Co. (Monrovia, Calif.) dreams up whole new lenses and camera systems for recon vehicles in space.

—Moog Servocontrols Inc. (East Aurora, N.Y.) worked out a unique turnabout retro-rocket system for the tail-end-to landing maneuver of lunar craft.

—Gaining overnight fame, the G. T. Schjeldahl Co. (Northfield, Minn.) devised the paper-thin aluminized plastic skin for the Echo satelloon, still in orbit (and the firm has since passed from the small-busines to not-so-small category.)

If you are soon to seek your aerospace job, one other common denominator among the space small-fry may interest you. Almost without exception, more than 50% of their personnel are scientists, engineers and top-notch technicians. But the prize perhaps is the 2-year-old Rocket Research Corporation (Seattle, Wash.), in which the total staff of 12 includes 12 scientists and engineers—namely, 100%. With a nationwide average of 75% trained brainpower holding down its total jobs, the many thousands of lightweight, specialty corporations pack a heavyweight technological punch.

You may find your big job for life among the small firms which are astronautics aces counting as America's trumps in the space game.

13 : TECHNOLOGICAL TRENDS

(Future space jobs unknown today . . . What new careers will arise in your lifetime . . . Solar system exploration before the end of the century . . . Astro-research fields most likely to expand . . . Great LASER breakthrough . . . Power conversion field . . . NASA/MSFC post-lunar program . . . Advanced interplanetary technology to come . . . Moon factories, Mars outposts . . . Space repairmen . . . Bionics and "engineered humans" . . . Anti-gravity . . . Faster-than-light stellar ships . . . Project Star-search.)

When you go to work for Ford, General Motors or other auto-makers, you are reasonably certain that next year's car won't be radically altered. Body-style touches and a few inner engineering improvements may be incorporated, but the engine doesn't sud-denly change from a combustion motor to a cesium-ion drive, and the fuel remains a refined hydrocarbon instead of becoming un-symmetrical dimethyl hydrazine enriched with liquid dynamite additives.

But when you go to work for Convair, Rocketdyne or any other aerospace firm, you honestly will not know just *what* you'll be working on next year. Your firm's particular chemical rocket may go into the discard while an ion-drive contract moves in. Last year's headline-maker among boosters may be next year's relic in a Space Age museum.

A significant lesson is to be learned from this: *most of the prod-*

ucts or systems you will be working on ten years from now do not even exist today. The Atlas ICBM, born out of frantic missile competition in 1958, is already being phased out with the Atlas-F series in 1963, to be superseded by the more advanced Titan and Minuteman. The Mercury Program has had its brief moment of glory sandwiched between the first flight in 1961 and the last in 1963.

But even faster and more revolutionary technological leaps into space are in destiny's timetable. Your *best* job ahead may be in a field barely starting today or not even named. To be shrewd, plan your future career in an *unknown* area of today. This sounds like double-talk, but it isn't. These future areas are already casting their huge shadows here into the past as exploratory studies or experimental R&D take place.

Experts have listed the following top-priority fields, virtually unexplored, that are almost certain to expand like an inflating balloon:

—LASER applications.
—Solid-state physics and molecular miniaturization ("molectronics").
—Plasma physics and hydromagnetics.
—Cryogenic magnets and superconductivity.
—Pyrogenic metallurgy (for re-entry at hypersonic velocity).
—"Rare" metal applications (beryllium, tantalum, niobium, zirconium, titanium, molybdenum, others).
—Space environment simulators (for manned or unmanned purposes).
—Bio-astronautics (manned spaceflight).
—Power conversion via fuel-cells, thermionics and solar energy.
—Compact computers approaching human capacity.
—Exobiology (extraterrestrial life).
—Interplanetary ecology (outposts on other worlds).
—Gravitics (gravitational phenomena).

Some of these, and other exotic technologies of the future, are worth exploring in further detail.

MIGHTY LASER

Trees were budding that day in May, 1960, under a warm spring sun. In the large apparatus-crammed laboratory, the scientist turned the rheostat up slowly, feeding more milliwatts into the system as a light-beam struck a tiny ruby crystal under electrical excitation. Another pencil-thin beam of crimson light, much more intense, sprang from the ruby to a meter. The scientist gasped audibly.

The target-meter's needle had swung to a full watt. He had delivered power—or pure light-energy—across a room without wires and beefed it up without discernible loss. How? By funneling ordinary light through a crystal which reconverted the mixed pulsations into one "pure" wave length of "coherent" light. And this allowed him to pack in any amount of power and "beam" it through the air.

The scientist would have had a right to yell "Eureka!" His LASER (Light Amplification by Stimulated Emission of Radiation) soon proved to be the breakthrough of the decade, the biggest thing since atomic power. It was a big-brass type of discovery, ranking with the electric light, radio, TV and the space rocket.

This milestone occurred in a laboratory of Hughes Aircraft in Malibu, California. The researcher's name is Dr. Theodore Maiman, Senior Staff Scientist. His achievement (based on the prior theory of two Bell Telephone Laboratory scientists) struck fire all over America and around the world.

Today over 400 corporations and many first-rank laboratories—Bell Telephone, MIT's Lincoln Lab, USAF's Rome Center—are researching this electronics wonder-child. Reports have flown in of amazing results as raw energy was hurled to any desired target—killing mice at ten feet, slicing razor blades neatly apart, drilling microscopic holes in diamonds, blasting through armor plate, shining a spotlight on the moon. The future implications are tremendous, especially since the LASER is a science virtuoso's dream, able to play an astonishing repertory of technological tunes: as a carrier-wave for long-distance space communications, for spot-welding of refractory tungsten, and for knifeless surgery with a beam-focused "scalpel," to name only three widely divergent suggestions.

Developed into a high-powered megawatt projector, the LASER may well be the long-sought anti-missile weapon astonishingly similar to a Buck Rogersish "ray-gun" that will be able to blast a flock of ICBM's high in their space arch. Or it may even be the key that unlocks thermonuclear energy, by squirting billion-degree quanta into fusionable plasma and starting its atomic fusion fire.

Perhaps the most important possibility is that of hanging great sun-gathering mirrors in orbit and beaming this photon flood down to earth via LASER "power lines," thus tapping the greatest pool of energy known. Solar energy pouring down all over earth each *year* is equal to *all* the atomic power locked up in our total uranium ore reserves on earth, mined or unmined.

Other breakthroughs comparable to the LASER, or not far below it, have occurred in this general area of *power conversion*. They have generally come out of advanced space laboratories, which need exotic power-plant hardware to aid man's future invasion of the universe:

Fuel-Cell.   Direct conversion of *heat* (of burning fuels) to electricity without use of the mechanical armature of present-day magnetic-field dynamos.

Thermo-electric Generator.   Using another form of heat from high-temperature plasmas (thin ion "gases") to gain simple, direct electrical current with lightweight portable equipment.

Thermionics.   "Cold" electricity from "atomic batteries" (successfully used in Transit IV-A satellite), in which a speck of radioactive isotope supplies atomic radiation to create a flow of electrons (electricity).

AFTER THE MOON—WHAT?

Technological prodigies of the above calibre will undoubtedly materialize in your lifetime. In fact, *you* may uncover one of them, if you go into the business of mystery conversion—turning the un-known into the known. There are vast areas in which you can profitably spend a lifetime tackling tantalizing challenges. Those lying ahead in our space program are being parceled out to various aerospace-industry labs by NASA. Most of these far-out concepts are being methodically plotted at Marshall Space Flight Center, simply because the man in charge there—Dr. Wernher von Braun —has the jump on almost every man on earth in working out the grand strategy of space invasion. Way back at Peenemünde, out of Hitler's earshot, he and his visionary crew talked of things beyond the V-2, of a far greater "war"—the conquest of space.

Von Braun's program will go far beyond the lunar-landing Apollo Project, which looms so large to us today yet is a mere stepping-stone to the truly great ventures among the planets. For this post-1970 campaign, the initial feasibility studies must start today. The following can be your guide to what fields of post-lunar hardware technology will be in progress during *most* of your life span. After all, we will dispose of the moon within 4 or 5 years, 7 at most. You will barely be starting your career, warming up for the big jobs ahead. And we mean BIG.

Mission	Contracting Firms (Feasibility studies only, no hardware)	
Post-Nova Launch Vehicles—nuclear rockets, ion-drives, electric propulsion.	Douglas: GD/A: Rand:*	$ 73,000 85,000 86,000
Sea-Launched Space Vehicles—mobile "floating pad" capable of launching rocket with payload of one million pounds into earth-orbit.	Aerojet-General:	$130,000
Reusable Boosters—systems for recovery of Saturn C-5 class and larger boosters, whose payloads range from 50 to 100 tons into orbit.	Boeing: NAA:	$150,000 142,000
Solid-Fueled Saturn C-1—replacing H-1 liquid-fueled engines with solid-propellant cluster of same thrust and payload capability.	Lockheed:	$197,000
Solid Nova—substitution of solid-fuel propulsion for liquid (thrust over 10 million pounds, payload up to 250 tons).	Boeing:	$139,000

Low-Thrust Space Transportation Systems—start-from-orbit craft with multi-ton payloads, using ion-drive or other nuclear/electric propulsion, to serve as supply ships to manned interplanetary outposts.	Rand:	$ 70,000
Planetary Trajectory Manual—list of all possible orbit-transfer paths (earth to planets) with guidance, energy and velocity requirements for Mars, Venus and Mercury.	Lockheed:	$ 85,500
Earth Orbital Operations Handbook & Lunar Flight Manual—orbit rendezvous techniques for manned space stations, plus take-offs from there to moon.	Martin-Marietta:	$150,000
Advanced Lunar Transportation—both Saturn C-5 with added stages and Saturn/nuclear vehicle, to ferry supplies to lunar camp on regular basis.	Not announced	
Early Manned Planetary Mission—to Mars or Venus in nuclear rockets first lifted into orbit by chemical Saturn C-5 or Nova.	Not announced	

* Rand Research Corporation (non-profit).

You can see from these studies, to which brilliant minds are being assigned in growing numbers, that just when you are hitting your stride after 1970 (if you are a student today), moon trips will be old hat and you'll be in the middle of a gigantic wave of exploration covering the solar system, world by world.

But even that is only the beginning. Aerojet-General and GD/Astronautics reveal they are doing company-funded studies of extraterrestrial *colonies*. Dr. Theodore B. Taylor, senior research administrator for GD/A, predicts cities on the moon by 1980, in which busy plants will manufacture spaceship fuels from native raw materials.

You will hear many ultra-conservative scientists snort at these visions. But just remember one thing—they are the same ones who, in their youth, shocked *their* conservative elders by exuberantly predicting aircraft that would fly across oceans and even achieve supersonic speed—and who made these dreams come true. But oddly enough, the science "rebels" of one generation are the "reactionaries" of the next. Growing older, they see too much of the

forest of hardware problems to take note of the vigorous technological trees springing up. Their tired minds, in too many cases,
simply lose the flexibility of scientific imagination, which first sets
the goal and tackles the difficulties later, instead of the reverse.

The von Brauns and the Krafft Ehrickes avoid such pioneering
petrification of their minds and join in Dr. Taylor's belief, that "a
considerably larger scale of activity in space should be predicted
for the years between now and the end of the century than has
been generally suggested. Regularly scheduled trips to the moon
will carry hundreds of passengers per year by 1980 . . . Mars will
have been thoroughly explored . . . The two small Mars satellites
will be permanent refueling stations and observatories. In 1990, the
human race will have explored the solar system in detail . . ."

Lynn A. Hannum, of Aerojet-General's Advanced Space Research Division, agrees, and also foresees strange Orbiting Factories
above earth, whose special vacuum-technology products will be
labeled "Made in Outer Space Only . . . You Can't Duplicate it
on Earth." And as dozens upon hundreds of "working satellites"
go up—advanced talking Telstars, Tiros weather-watchers, earth-
mapping GeoSats, ship-guiding Transits—a host of space repairmen
will be needed, riding up in rocket ferries to service and maintain
those expensive and vital space aids in the sky.

You might become one of those space specialists. No ordinary
workman will be able to handle the perplexities of zero-g phenomena without understanding basic physics, or fix a LASER
power-projector without years of prior engineering practice on
earth. Scientists and engineers will most likely, NASA itself predicts, make up the *majority* of all space explorers who first set foot
on other worlds and extend our astronautics links throughout the
solar system.

Therefore, if the age-old red blood of adventure lurks within
you, then think seriously of your career in *space*—literally.

Now that we've disposed of the routine, let us get to the really
unorthodox. There are hints, from the speculations of men with
faraway gleams in their eyes, that some truly staggering surprises

will greet you more than once during the span of your Space Age life. Take the following with the same grains of salt that were sprinkled over the wild dream of atomic energy by 98.67% of *all scientists*—prior to 1940.

HOMO SPACIENS

A new breed of *Astro-Men* who supersede the current astronauts may be adapted through scientific means to space conditions that would kill ordinary earthlings. Like super-mutants, they will be impervious to sub-zero cold or metal-melting heat, subsist for year-long interplanetary trips on a hundred drawn breaths of oxygen, ignore solar death-ray storms and think with the ultra-speed precision of a computer.

These radical metabolic alterations in spacemen are being investigated by the new science of *organo-bionics*, which means combining miniaturized electronic gadgetry with the processes of the human body. If this seems "unnatural," then let us be consistent and remove gold fillings from our teeth, or artificial tantalum rods in place of smashed leg-bones, or platinum skull-plates which keep us alive and happy after bad accidents.

The aim of these "mad scientists" is quite sane, in fact, humane: to prevent human suffering during the ordeals of space travel in an alien environment. The organo-bionicists plan to insert pea-sized sensors, heart-timers, gland-inhibitors and other such physiological monitors beneath the human skin. Suitable currents and charges will then put the spacemen into virtual "suspended animation," which is simply another term for the very natural hibernation technique animals have used for ages against *their* adverse "space" (winter) conditions.

The gadgets will be turned off on arrival at an earth-like environment, such as a Martian colony, and the electronic zombie will once again return to his normal, smiling self. None of this is any wilder than one Space Age technique already being used within the human body to save lives—a tiny electronic timer, borrowed from a rocket's guidance system, that sends gentle electrical pulses into ailing hearts and keeps them pumping rhythmically. You may pass one of these ex-heart patients any day. Thousands of them now walking around in active life would be in a hospital bed, or a coffin, except for their "space timers." They do not feel "unnatural" with that miniaturized mercy device buried in their flesh. They feel just fine.

But space medicine will gradually break down any such lingering prejudices people may have against "tampering" with the human body. Organo-bionics is apt to grow rapidly into an immense area of "human engineering," in order to make the conquest of space possible in the fastest and best way for frail humans. And it will likely, as space-research fall-out, make the conquest of earthly ailments possible too and benefit human health in revolutionary ways.

PUSHBUTTON GRAVITY

The neutrino is a sub-atomic particle of no mass—namely, *no weight*. Impervious to gravity forces that matter exerts, the neutrino can therefore plunge blithely through some 4000 feet of solid lead before it stops. Recently, the anti-neutrino was discovered, a sort of "mirror image" twin that displays a repulsion to normal matter, in short, *anti-gravity*.

This stunning discovery led to an anonymous announcement from behind the USAF's space technology security curtain: "There is no reason to believe that this massless particle cannot penetrate gravitic wave fields and achieve full escape. This principle might also be applied to matter in bulk."

Two shockers lie in that candid bombshell: one, the implication that scientists are on the track of gravity *waves*; two, that the anti-neutrino's ability to escape these waves might be extended to large objects. Both add up to the old science-fiction theme of anti-gravity forces which release objects from earth's gravitational pull or are even repelled. From this could come repulsion propulsion, to waft heavy space ships away from earth like thistledown.

Do not laugh too soon. It is now on record that an unknown number of top scientists and organized laboratories—both in America and in Russia—have been working on "para-gravity phenomena" for some two years. Their guarded statements in obscurese (Washington's official language for classified projects) cannot obscure the central motive—to find out if gravity operates in *waves* (like all electromagnetic phenomena), and if so, can it be "shielded off" just as ironware will stop radio-waves.

You can add up the rest yourself. In fact, you *may* be adding up the rest in your future career and making the hardware to find out if it works. Any form of anti-gravity force applicable to spacecraft —say, reducing its weight (or earth's gravity pull) by 50%—would obviously revolutionize space technology and automatically double all payloads we launch.

To be objective, anti-gravity may be just a dream. Maybe not. Time will tell—but not any "expert" today, or even a group of Nobel Prize winners a hundred strong. Nobody today can state flatly that a gravity breakthrough is impossible. That would take futuristic insight, which no feasibility study has yet proved exists within human skulls.

TRANS-PHOTONIC DRIVE

Is faster-than-light spaceship propulsion possible? Despite the outraged screams of those who revere Relativity, there is now honest doubt that the great Einsteinian absolute—that the speed of light in vacuo is some 186,270 miles per second—is the final velocity limit in the universe, forever confining our craft to sub-light speeds.

Cracks are appearing in this once-impregnable foundation, according to accumulating reports from various scientific quarters. Dr. Martin Ruderfer (the President of Dimensions, Inc., a research firm) suspects from slight aberrations in predicted eclipses of Jupiter's moons that the speed of light between that planet and earth may not be the same both ways. We use only an *average* value in all celestial cases, which immediately implies that light goes *faster* than our accepted value one way, slower the other. The slower is as significant as the faster. It may account for the famed "red shift" in distant galactic spectra and destroy the theory of the expanding universe.

Another startling phenomenon—this time backed experimentally —was produced by an East Indian scientist who made cyclotron-accelerated electrons move faster than light through water. Even though this value did not exceed light's vacuum speed, the mere surprise that something can beat light in *any* medium is revolutionary. The possibility immediately opens that such electrons might beat "unbeatable" light in a space race too.

This indeed would blow the lid off Relativity, even more than anti-gravity. Gone would be the dirgelike dogma of scientists who inflexibly state (though Einstein himself never did) that because of light-speed limitations and the vast gulfs between stars, mankind will *never* make the long trips of 10 or 100 or 1000 years necessary for reaching other remote civilizations. They forget that this conclusion is based on Einstein's *theory*. Theories change or need modification. The stellarship *Photon Greyhound,* using its anti-solar-gravity drive at full repulso-thrust to reach its maximum velocity of 1000 light-speeds, would bring a large part of the gigantic space realm within visiting range. It *can* happen here in the age of space revolution, even to the sacred cow of the Relativists.

As the somewhat paradoxical adage has it, the only unchanging thing in the universe is that all things *must* change. Human civilization is on the verge of stupendous new scientific revelations that will probably topple 95% of today's orthodox thinking, leading us to technological insights as far beyond us as we are beyond the pre-wheel cavemen. If, in this process, classic Einsteinian theory springs leaks, obviously the staunch believers in it will have to accept the necessary modifications, which may (or may not) include removal of light-speed as a limitation on galactic space travel.

Or, if you find this dream shattered by inviolable Einsteinian gravity and/or unsurpassable light-speed, join Project Starsearch, headed by such world-famed astronomers and astro-physicists as

Dr. Otto Struve, Dr. Gerard Kuiper, Dr. Thomas Gold and Dr. Fred Hoyle. They estimate that at least a million (more likely a billion) "twin earths," or at least habitable planets, exist in our own galaxy alone. Perhaps a dozen are within 25 light-years and a thousand within 100 light-years, in the expanding bowl of star-filled space around our tiny world.

By radio-telescope, MASER-sensitive receivers, even infra-red and ultra-violet detectors tuning through unconventional communications bands, a large group of these eager cosmologists plan (hopefully, with eventual government support) to scan systematically the star-studded skies with their ultra-hearing aids, waiting breathlessly for that first electrifying coded whisper that cannot come from random nature but only from apparatus created by rational minds.

Is their venture a hopeless long shot? Here is a paraphrased abridgment of a sober yet unqualified statement they issued: "Taking into account the swift advancement of our detection apparatus into multi-parsec ranges, and using the dependable law of averages as to how many other worlds must have technology exceeding earth's, there is little doubt that some form of signals of outer-space origin, from some race of thinking creatures, will be picked up *within the next ten years.*"

You that graduate soon can hurry to become one of the staff of this Galactic Listening Post. You might just happen to be the first human to hear that non-human message. Somebody has to be first, and there are literally hundreds of "firsts" waiting to be fulfilled, simply because almost everything in astronautics is something new under this sun—or any sun.

14 : CONCEPT TO COUNTDOWN

(What a space job is like . . . Interview with RCA satellite engineer . . . Sidney Sternberg's story of Project TIROS . . . The TIROS team . . . Concept and design of first Weather Satellite . . . Success of TIROS I and following series . . . Engineering details of instrumentation . . . Technological snags that were overcome . . . Follow-on Nimbus and Aeros Projects . . . Personal rewards to Sternberg from his space career.)

Just what is it like to be a member of America's space team? What kind of work do you actually do when you are on the job, whatever that job may be? What technological hazards arise and how are they solved? How does a team of astronautics experts turn a bare unproven idea into a shining triumph whirling high in space and serving mankind?

To give you some idea of what your daily life may be in your space career, here is a true-life parallel from a direct interview. Project yourself into the place of the man we will present. You are now Sidney Sternberg, space engineer . . .

Sidney Sternberg, chief engineer of the Astro-Electronics Division of RCA, is a young man in a very young industry. Indeed, the TIROS satellites, built under his supervision, represent the space industry's first "practical" result. Today, when the Weather Bureau calculates its forecasts, part of the data on which they base their predictions come from the latest TIROS, circling the globe every 90 minutes in a 400- to 500-mile-high orbital path.

The original concept of TIROS evolved during a series of meetings called by the Defense Department's ARPA (Advanced Research Projects Agency) early in 1958, to discuss the possibility of a TV-equipped satellite for obtaining detailed pictures of the earth's cloud-cover.

Vanguard II and other earlier satellites had demonstrated the technical and engineering feasibility of this kind of project. The U. S. Weather Bureau felt that the information from cloud-cover pictures would be invaluable toward gaining a more complete understanding of weather phenomena.

Present at these meetings were representatives of the Weather Bureau, NASA, the USAF and the Cambridge Weather Research Center. RCA, which had already done pioneer work on TV satellites, was also invited to attend.

At these conferences the basic "philosophy" (trade term for *over-all procedure*) was thrashed out. A Thor-Able booster was decided on for lofting the payload into orbit, and specific performance goals were established for the instrument package itself. RCA was awarded the contract for designing and building the system. They, in turn, assigned the job to Sidney Sternberg in the newly organized Astro-Electronics Division.

"We knew the limits imposed on size and weight by the capability of the Thor-Able booster," Mr. Sternberg recalls. "We also knew what the finished satellite was expected to do. Beyond that, we were on our own—limited only by our talents and imagination."

One might expect the desk of the busy director of this kind of technical installation to be cluttered with papers and abstruse formulas, but Sternberg's desk is almost bare. Outside of the purely functional telephone, an intercom speaker and a pen holder, the only other things on his desk are a model of the TIROS and pictures of his children. These represent the central interests in his life—his family and his work.

The office where Sternberg works is simple, almost spartan.

There is a carpet on the floor, but you don't sink into it up to your ankles. A large desk, a table running along one wall, a few chairs and a cabinet complete the furnishings. Everything is orderly, functional and completely without pretension.

The same modesty and orderliness reflected in his office are also evident when you speak to Sidney Sternberg. Of medium height, with black, curling hair beginning to gray at the temples, he speaks easily, and quickly reveals a sharp intelligence and a warm humor.

GAINING KNOW-HOW

Since he joined RCA in 1951, this brilliant engineer has been involved in much of the pioneering effort on the development of electronic-scan space systems. He was appointed Chief Engineer of the Astro-Electronics Division when it was first organized in 1958. Under his direction, this RCA division is conducting study, development and construction programs on earth-satellite systems, space probes, rocket-propulsion systems, secondary (payload) power systems and specialized satellite components. Two notable achievements under his technical management, besides TIROS, have been the design of the ground and satellite-borne communications equipment for Project Score and the entire satellite-tracking and ground station equipment for all meteorological satellites.

Mr. Sternberg is a native of New York City, where he grew up in the Woodlawn section of the Bronx. He attended De Witt Clinton High School and received his B.S. degree in physics from the City College of New York in 1943. After graduating from CCNY, he served for three years with the U. S. Navy as a radar officer.

"Except for six months I spent in the Philippines on radar maintenance," Mr. Sternberg reveals, "most of my time in the Navy was spent in schools. I attended Northwestern University, Harvard and M.I.T., where I took Navy courses in radar and electronics."

After his discharge from the Navy in 1946, Sidney Sternberg

went back to school. He attended New York University, where he received his master's degree in electronic engineering, while working as a civilian employee with the Special Devices Division of the Navy's Bureau of Aeronautics. He worked under Admiral Louis DeFlores on the Navy computer program, developing such equipment as the Link Trainer, which can electronically simulate the actual conditions that occur in flight, and the computers for the radarscopes of Project Typhoon (airborne radar reconnaissance of storms at sea).

"At that time, there weren't too many people available with a background of electronic pulse experience—experience we gained working with Navy radar," Mr. Sternberg told us. "This was still a virgin field then and needed a good deal of basic research for its development."

It was while working with the Navy Bureau that Sidney Sternberg first came in contact with RCA. A development contract was awarded to the RCA Laboratories (Princeton, New Jersey) for exploring the project. After working with their engineers, Mr. Sternberg was invited to join the RCA Laboratory, which he did in 1951. He continued working in research programs typified by the Typhoon computer and the Dynamics Systems Synthesizer. He received the RCA Laboratories Research Achievement Award first in 1953 and again in 1955, and has several assigned U. S. patents.

"Our Dynamic System Synthesizer was able to simulate the actual conditions of a missile flight," Sternberg said, "which enabled our mathematicians to formulate the total equations for such an operation."

In 1955, when a previous study on the use of television for space reconnaissance systems was broadened into research on an over-all system, Sternberg was put in charge of the Information Systems portion. He became project manager of a feasibility study which led to design of a television reconnaissance satellite system. Out of this work, the concept of the TIROS emerged.

Sternberg has a gifted crew under him today. Edwin Goldberg, a distinguished gray-haired man with a slow drawl, is Project Manager of RCA's TIROS Satellite System project.

Mr. Goldberg joined the engineering staff of RCA in 1940 after receiving his M.S. degree in electrical engineering from the University of Texas. A specialist in development and design, Edwin Goldberg has 20 issued U.S. patents to his credit. When RCA organized the Astro-Electronics Division in 1958, he was named manager of the hardware-producing section.

Under the direction of Mr. Goldberg and his aide, Vernon L. Landon, the engineering staff of this RCA subsidiary designed and built what is acknowledged to be the *most successful* satellite to date. His viewpoint on how you make a satellite is interesting.

"It's like adding a room to a house without a blueprint," explained Edwin Goldberg, in his Texas drawl. "You add a little here, take off a little there, and for a while it looks like the darn thing will never fit. But it all works out and the new room, though it isn't exactly what you orginally planned, looks pretty good." He added with a wry grin, "Of course, building a satellite is just a bit more complicated."

Complicated is the understatement of the century. One of the first problems with TIROS proved to be basic to the ultimate functioning of the satellite as a stable camera platform. Explorer I, America's first satellite, had shown that *precession*—wobble that develops in a spinning body—would have to be overcome.

Precession had caused the pencil-shaped Explorer satellite to tumble about its center soon after it achieved orbit. Although this motion did not interfere critically with Explorer's job of collecting radiation data and meteor-hit counts, it would have been impossible to take pictures. Obviously, a different shape had to be devised.

"Before this problem was licked," said Vernon D. Landon, manager of the engineering staff, "we had to do a great deal of basic research into spin-dynamics."

This research, besides determining the shape of the satellite, also provided a theoretical basis for the ingenious anti-wobble and spin-control mechanism incorporated in the completed package.

At its separation from the third stage of the rocket carrier, the TIROS satellite spins at a rate of 120 rpm. To slow it down to the desired rate of 12 rpm, a "yo-yo" arrangement of two weights attached to the ends of cables is automatically extended outward from opposite sides of the vehicle. The weights "soak up" and dissipate the energy of the spin. When sufficient slowdown is achieved, the cables slip off open hooks and leave the package.

"They won't fall and hit somebody on the head," one of the engineers said reassuringly. "They're probably still in orbit. When they do enter the atmosphere, they'll burn up like meteors before reaching earth."

To overcome the wobble problem, one of the engineers hit upon a simple but effective arrangement of sliding weights. Moving along vertical rods in direct opposition to any wobbling motion, the weights absorb the energy of the motion and stabilize the satellite.

Both the "yo-yo" spin-rate reducer and the anti-wobble mechanism are activated by a special lift-off switch (a toggle that is automatically set when the satellite separates from the rocket). The lift-off switch triggers the firing of rocket squibs that release the weights.

But still another aspect of the spin-dynamics of the satellite was good for group headaches among the design demons. The magnetic field of earth, acting constantly on the orbiting vehicle, would eventually slow the spin below the optimum 12 rpm. To overcome this drag, small rockets were arranged around the rim of the satellite body. The rockets, fired on signal from the ground, were designed to boost the spin rate back to normal.

This boosting system was put to the test with TIROS I in early 1960. After 3 months in orbit, magnetic drag had slowed the spin rate to 9 rpm. The rockets were fired by ground control and functioned perfectly. They boosted the spin rate to 12.8 rpm and two more rockets remained for another boost.

"The problem of spin-dynamics, however, remained a constant king-sized headache—without a filter," said Goldberg. "The satellite had to be perfectly balanced and we juggled components endlessly before an optimum weight distribution was achieved."

OUTWITTING ZERO-G

Another limit was imposed by the mechanics of thermo-dynamics. In the nearly perfect vacuum at the altitude of the satellite, temperature (as such) does not exist. The heat of matter in a vacuum is determined solely by the radiant qualities of the material it is made of.

"The silicon in the solar cells, for example, has a high absorption ratio," explained Sternberg, since this was his research province. "That is, it absorbs more heat than it radiates. This property would cause the silicon cells to heat up beyond the point at which they lose their electric generating power. In order to compensate for this poor emission ratio, some means had to be devised to absorb the excess heat."

After exhaustive tests, Sternberg went on, it was found that a special type of glass placed on top of the silicon cells would do the job. The glass has a high emission ratio (it radiates more heat than it absorbs) and safely drains off the excess BTU's from the silicon cells without interfering with the electric generating function.

"These were the limits under which we would have to work," said Sternberg. "Each component of the payload would have to be a compromise between five factors: spin-dynamics, thermo-dynamics, minimal weight, electrical-power drain and maximal performance."

Within these rigid parameters (engineering boundaries), the main job of assembling the instrument system began. While the basic questions of configuration, weight and size were being answered, work on the components was proceeding simultaneously.

To coordinate the efforts of the various departments, periodic "progress conferences" were held. At these meetings the heads of various departments reported new developments and requirements, besides keeping the rest of the staff informed on the over-all progress of the satellite.

"Most of our conferences, however, were quite informal," explained Goldberg. "Max Mesner, for example, who is working out the TV camera design for the bird, might feel that he needs a wider frequency band. This means more power. So, he walks across the hall to the power department and spells out the new requirement. The power engineer goes over his figures and finds that he might possibly steal some amperes from the tape recorder. The two of them now stroll over to the tape engineers and sound them out. Yes, they can spare the power, they quickly agree (after a 5-hour argument). After it is cleared with the packaging people, who only dramatically tear their hair at the tight margin, the innovation is incorporated into the vehicle and duly reported at the next formal progress conference."

"This kind of friendly give-and-take, in our sound-proofed shout-room," Sidney Sternberg added with a straight face, "went on all through the construction of the satellite. The fact that the entire operation was under one roof facilitated the communication problem considerably. That is, when the engineers would talk to each other."

Many new facilities were added to the plant for the TIROS project. Special testing equipment was necessary to check out both the components (before they were fitted into the vehicle) and the completed satellite.

"Our bird would have to live in space," Goldberg pointed out. "To find out how it would act out there, we had to devise means of simulating the conditions in space as accurately as possible."

To achieve this goal, a complex of special testing equipment was installed in the plant. A vacuum chamber, designed to simulate the atmospheric conditions at an altitude of 400 miles and more, was installed; a magnetic coil was constructed to duplicate the magnetic field of earth; a gravity-stress machine, accurately duplicating every degree of stress TIROS would be subject to in launching, was built.

Each component was subjected to these testing devices. The parts that didn't stand up under the grueling strain were modified and redesigned until they did.

"Our storage batteries exploded," Goldberg admitted, "the first time we put them in a vacuum. The gas generated in the batteries expanded in the absence of atmosphere and could not be contained by the original casing."

The battery cases were completely redesigned. The structural housing was strengthened and the shape was modified. Finally, they held intact, even in the most perfect vacuum the testing apparatus could achieve. All this meant long weeks of labor in the laboratory, with a liberal schedule of sleep every other night without fail.

Vacuum testing also revealed bugs in the tape recorder and timing-clock mechanisms. The lubricants necessary for the proper functioning of these components vaporized in the vacuum. To solve this problem, the two parts were housed under hermetically sealed plastic domes where a pressure of ½ atmosphere is maintained.

"These two components, plus the camera shutters," said Sternberg, "are the only mechanical assemblies in the satellite. Our goal was sophisticated simplicity. We considered a paddle-wheel arrangement for the solar cells, for example, but discarded it in favor of the simpler cell-covered housing. With the more complicated paddle wheel, there would have been one other worrisome element that might not function properly."

However, for the most part, the absence of atmosphere and gravity proved to be a boon rather than a drawback.

"Space is really an excellent environment for delicate instruments," explained Sternberg. "There is no dust or atmosphere to interfere with their functions and no friction under gravity to wear them down. It's a nice quiet place where they can do their work undisturbed."

EYE IN THE SKY

One by one, each component of the vastly complex instrument system of TIROS I was tested, checked and rechecked before being put into place on the satellite's circular base. Gradually, TIROS took shape. And, as this work was forging ahead in the case of the first TIROS, another group of Astro-Electronics engineers simultaneously designed and developed the ground-control system.

"This gave us one more measure of control over the total design," said Sidney Sternberg. "It meant that, wherever possible, we could simplify the elements within the satellite complex at the expense of the ground-control system. Maintaining the equipment on the ground would be far easier than maintaining the satellite—400 or 500 miles above the earth."

This has been proved true with the half-dozen TIROS orbiters to date.

Finally, each satellite is fully assembled. But before it can be launched, it must be subjected to the same grueling tests that the individual components have been exposed to. The entire TIROS is placed within the vacuum chamber, where it remains for two weeks. Next, it goes on the gravity-shock machine and is subjected to repeated stress far greater than it will experience in actual launching. The reaction to the magnetic field is studied in the giant magnetic coil.

Functioning now as part of the whole, each component is exhaustively tested before it is finally checked out.

The satellite carries two cameras, identical in every respect except lens size. Both are about the size of a water-glass and employ a sturdy half-inch Vidicon pick-up tube.

One of the cameras has a wide-angle lens for viewing an area of cloud-cover nearly 800 miles on a side. The second camera, in earlier versions, had a narrow-angle lens for detailed viewing of a smaller area covering 10,000 square miles. This has been replaced in recent TIROS vehicles by a medium-angle lens viewing a larger area. The two cameras and their associated electronic devices operate independently, so that any failure in one will not interfere with the operation of the other camera's chain of sequences.

The cameras, in contrast to broadcast-type television equipment, are designed to take still pictures, exposed by means of electrically operated shutters. Shutter speed on each camera is 1.5 milliseconds; lens openings are F1.5 and F1.8 on the wide-angle, and narrow- or medium-angle cameras, respectively.

Associated with each camera is a specially designed tape recorder for storing the TV pictures, to be played back on command from the ground. Each recorder is housed in a pressurized dome. Capacity is 400 feet of specially developed half-inch tape in each recorder, moving at 50 inches per second during recording and playback. As the pictures are played back for transmission, the tape is automatically erased, preparing it for further recording on the next orbit.

Besides recording the pictures, both tape machines also imprint coded signals indicating the rotational position of the satellite, as a time reference to help identify the pictures when received on the ground.

Both camera-tape recorder systems have their own transmitter: a 2-watt FM unit operating in the 200-megacycle band. Each transmitter weighs one pound and occupies a space measuring 6 by 3.75 by 1.75 inches. The transmitters operate only on command from earth control while the satellite is within range of the ground stations. Except for one pre-amp tube, the entire electronic system is transistorized.

As the satellite comes over the horizon within range, a ground-station signal activates the transmitting mechanism and either or both of the tape recorders starts the playback of the pictures taken during prior orbit.

The TV-camera/transmitting chain can also be activated to transmit pictures directly to the ground without storing them on tape. In this case, the TV systems are controlled directly by signals from earth, bypassing the sequence timing of the electronic clocks and the storage function of the tape recorder.

DEBUT, TIROS I

As an aid in interpreting the pictures of cloud-patterns, two reference systems built into the vehicle show its direction and location in reference to earth's surface—or how the orbiting satellite "slants."

One reference establishes the attitude (position relative to an earth-center perpendicular). Small infra-red detectors mounted in the vehicle sense the heat radiation from the earth as the satellite spins on its axis at 12 rpm. As the infra-red scanner sweeps across the earth, an electric pulse is developed. Signals marking the beginning and end of the pulse—corresponding to the two edges of the earth—are sent to the ground station, where they are timed and recorded on a punched-tape recorder. The information is then processed to give an accurate account of the spin-axis attitude of the satellite at the time the picture was taken.

The second earth-orientation reference is the north direction. This is obtained by an arrangement of recessed solar cells at nine points around the sides of the vehicle-housing. Each cell, covered with a plate containing a narrow slit, "sees" the sun for a brief instant through the slit as the satellite rotates, and produces an electrical pulse. The pulses differ from one cell to the next as the angle of the sun changes. Transmitted to the ground with each picture, these pulses tell the direction of the sun (and therefore the earth and its north pole) in relation to the satellite at the instant of picture-taking.

"We put the completed bird through the most exhaustive tests we could devise," said Goldberg. "We wanted to be sure every part functioned before we okayed the satellite for launching."

After everything was checked out, and performance standards satisfied the critical engineering inspection staff of Astro-Electronics Division, TIROS was ready to fly for the first time, back in 1960. Five satellites in all had been constructed; three of them were shipped to Cape Canaveral for launching—two for back-up reserve. Before launching, some NASA people wondered whether the satellite would really work, once in orbit.

"You just get it up there," Sidney Sternberg told them. "That's the main problem now. Don't worry about TIROS—it'll work fine."

He was wrong. It worked *magnificently*, far better than anyone expected, sending back a record of cloud pictures that exceeded official expectations. The unqualified success of TIROS justified the confidence of the RCA engineers who made it.

"We weren't too surprised," Sidney Sternberg told this reporter. "Those of us who were close to TIROS knew the kind of work and effort that went into our bird. We would have been more surprised if it hadn't done us proud."

LIFE WITH SPACE

This ace of space engineers lives with his wife and three children, two boys and a girl, in Princeton, New Jersey, near the brand-new RCA Astro-Electronics Center. Like so many other scientists and engineers, he is completely engrossed in his work.

"The demands of my job," he said, "take up almost all of my energy and interest. When I do have some free time, I like to spend it with my family."

A good portion—perhaps too much—of Sternberg's time is spent traveling.

"Sometimes it seems I do most of my thinking work in an airplane." He shrugged his shoulders resignedly. "Our plant is in New Jersey; the Thor booster used to launch the TIROS is made in California; the blast-off takes place in Florida; our ground installations and our engineers are strung out in a line that reaches from

Puerto Rico to Hawaii; and our subcontractors who make component parts for TIROS are spread all over America."

He added matter-of-factly, "My job is to coordinate all the far-flung facets of our operation. It's simply a matter of communications. I have to see and talk with all of these people, there's no other way of doing it."

Sidney Sternberg has a dream that will one day solve this most pressing and time-consuming problem. He envisions a time when all of the people he deals with are hooked up on a closed-circuit TV system. He would have a million cycles of bandwidth at his disposal, and when he wished to speak with any of them, he would turn a knob and they'd be right there in his office.

"Until that kind of system is feasible," he grinned, "I guess I'll just have to keep flying."

At present, the Astro-Electronics Division is preparing an indefinite series of further TIROS vehicles which NASA feels will be useful up to 1965 if not beyond. The design specifications for this series are almost identical with those of TIROS I, so perfect is its basic system.

The next step in the weather reconnaissance program is the Nimbus, parts of which have been under development by various companies, including RCA (and GE/MSD as prime), for NASA. Nimbus launches are scheduled from 1963 on through the rest of this decade.

"We worked out TV-type Nimbus sensors, including a new electro-static tape camera," Sternberg said. "The whole Nimbus system is a flexible one, with interchangeable components that can be utilized for a wide variety of experiments. Beyond Nimbus will be an entire series of advanced WeathSats called *Aeros* that will open even broader vistas of scientific knowledge, and usher in amazingly accurate weather forecasting."

Unlike many, Mr. Sternberg does not despair over the progress America is making in the Space Age.

"Aside from being a little late," he believes, "our over-all program is moving ahead fast. American satellites and probes are collecting much valuable scientific data and we've been able to develop operational space systems that are already beginning to benefit the public."

These advances, Sternberg points out, represent a technological ability second to none in many areas. And as bigger and more powerful boosters become available to us, he said, this technological lead will become even more significant in the Space Race.

"Science," Sternberg says, "has become an instrument of national policy. The result has been a tremendous boost to the entire spectrum of scientific effort, on earth as well as above."

Nowhere is this scientific effort more in evidence than in Sternberg's Astro-Electronics Division. In the future, other advanced vehicles, designed and built under his direction, will advance the frontiers of science far beyond its present limits.

Asked finally if his work was hard, Sternberg looked surprised. "Why, I never really thought about it," he said slowly, then shook his head and answered with a question of his own: "What's hard about doing something you *like?* Something that gives you a fresh feeling of excitement every morning, and that contented feeling each night of having done a good job? Hard or not, what could be better?"

What could, indeed?

15 : WHY SPACE?

(*Review of America's space motives . . . Challenge of technology explosion . . . Soviet mental mobilization . . . Untouched USA brain bonanza . . . Anti-intelligence prejudice . . . Myth of "evil" science . . . Simon Newcombs of space . . . Space surprises ahead . . . Your space career by 1975.*)

In this final wrap-up chapter, if "why space?" cannot be clearly answered for you, perhaps "why *not* space?" can be. Some conjectures in this book have probably come to controversial conclusions.

One thing is clear, however. It is not a false alarm that America is at a crossroads. To think so is to feed the spirit of complacency, which prevents action. Just as 1776, 1861 and 1941 represented critical periods in our nation's history, with relatively untrying times between, so today we are faced with another crisis. Perhaps the greatest of them all.

Reaching into all international as well as national facets of human life, a Science Revolution is shaking the world. The atomic bomb and the space rocket together, occurring almost simultaneously in historical terms, have truly blasted open a Pandora's box of change in all things.

Change for the better or the worse? That is precisely the challenge facing *you*, as the full impact of galloping technology gathers speed in your lifetime ahead. Whatever the cause of this science technology explosion, it is here in front of us—a renaissance of naked brainpower such as the human race has never known before. Over 90% of all the scientists who ever lived are alive today. The U.S. budget has allocated more to research since 1950 than in the previous two centuries of its existence. Elsewhere in the world— Europe, Russia, Japan, India—the same floodtide of organized assault on nature's secrets is evident. Even the young and "emerging" nations of the world, small and undeveloped, are building completely modern laboratories and staffing them with native scientists.

What the future implications and consequences of the Laboratory Explosion will be are as difficult to assess as its cause. This book makes no claim to superior foresight or precognition. But some things seem to stand out clearly in the harsh glare cast over the world by the blazing mental fires that have been lit. Let us get a final "fix" on the reefs lying in America's path and chart the methods necessary to avoid shipwreck. For your sake. Your generation will be at the helm.

MENTAL MOBILIZATION

First of all, is the Soviet threat real? Some believe that the Cold War will fade away and Russia will eventually cooperate with us fully in space, as they have begun to do in a small way. This could happen, and we can all hope that it will. Even if it does, this is no reason for America to stop forging full steam ahead into the frontiers of science in general and space in particular. The goal of new knowledge is worthwhile for its own sake.

However, many find it unrealistic, in terms of international events in recent years, to expect the all-round rivalry between the USA and the USSR to cease—in earthly affairs or in space—in the foreseeable future. An increasing number of leaders in education, government and industry warn that America has fallen dangerously behind in turning out the technological brainware needed for the struggle ahead.

In contrast to our decreasing number of graduating engineers, and the slow rise in number of science grads, the Russians have for years conscripted new mindpower, sparing no effort or expense. With only half the wealth of the USA—as an official report of the National Science Foundation reveals—the USSR matches our education budget, ruble for dollar.

"There must be no misunderstanding or underestimation of the Soviet scientific and technical manpower buildup," states the NSF report bluntly. "It has [already] become the principal source of Communist strength . . ."

The chief compiler of the NSF report, Nicholas DeWitt, goes on emphatically: "Soviet leaders firmly believe that the competition between capitalist democracies and the communist world will be decided in the field of science and technology . . ."

Switching to an industrial voice, Thomas J. Bannon (President, Western Gear Corporation) stated in relation to the Russian drive: "The challenge of this dangerous time must be met, not only by the weight of manpower as in the past, but by the weight of mindpower. It is the intelligence of the new generation, honed to a fine edge on the stone of learning, that will decide whether our manner of belief, and our manner of life, will survive."

Out of the Soviet harvest of brainpower—which was sown years ago—have come the Sputniks, Luniks and Vostoks that startled the world. And just when America's space effort is now gathering momentum, it faces a mindpower gap which can be more crippling than the notorious booster gap.

Falling behind in the brain race will be far worse than lagging in the space race, because we are dealing with *human* hardware. We can no more speed up the training of a youngster's brain than we can miniaturize his body. The memory banks of human computers cannot be filled a thousand times faster with the flick of an electronic switch. No new technique can appreciably shorten the average sixteen years of R&D—from first grade through college— that produces the operational mental machinery of a living scientist or engineer.

We can match the growing Soviet brainthrust only by launching the present generation of American youth into scientific schooling in doubled or tripled numbers. There is no other short cut for finding our badly underused brain bonanza, and processing it into mental gold.

SPACE SUPERSTITIONS

Assuming our educational budget is soon beefed-up properly, another problem remains, which only *you*, the students, can solve.

This goes back to that school poll and *your* attitude (collectively) of being *anti-intellectual*, or at best non-aligned and therefore indifferent. This is Stone Age prejudice you must overcome. You have fresh, young minds. You can shake off parental prejudices against using the supreme tool of intelligence. If our society opposes this, it is the same as picketing against progress, or refusing to bore into a mountain made of 24-carat gold.

Somehow, you must also find the *motivation* for enlisting in the shock troops of science and engineering America needs. You must reject worn-out values of pre-space times as to what "success" in life is. Your technically talented brain (if such it is) will be wasted in researching new stock-market systems, or engineering high-pressure sales campaigns, or achieving breakthroughs in banking.

None of this will make your generation a match for the legions of trained Soviet mindpower in 1975, who will clobber you and your future America through advanced science, no matter how many color TV sets, sports cars or jukeboxes the nation has.

If you are to grow up as a mentally mature generation, there is another American folktale you must reject—that science "causes" all the troubles in the world, by spawning its "dangerous" discoveries, notably the nuclear bomb. Let us ask some pertinent questions:

—Does a scientific discovery turn itself into an "evil" weapon, or is this brought about by calculating men and political forces?

—Do wars start in arsenals and bomb stockpiles—or in mis-guided human hearts?

—Would destroying all nuclear bombs insure world peace, if cannon and tanks were left? And if these and all other civil-ized weapons were banned, would that prevent unenlightened men from using clubs, fists or teeth on each other?

An equally serious barrier to space science, specifically, is em-bodied in the naive question still being asked by the American people—"Is it really *worth* spending all that money for a trip to the moon?"

Upon this parallel we can base an equally sensible (or non-sen-sible) series of questions, going back to various periods of history:

—Is this pioneering solo flight across the Atlantic necessary, Lindbergh?

—Are your royal jewels worth wasting on that crazy trip of Co-lumbus's across the ocean's edge, Queen Isabella?

—Is it worth swimming across that swift river, Og, to find out if another tribe actually built that fire we see?

SIMON NEWCOMB RIDES AGAIN

But if the layman has a one-legged imagination, what can we say of the professional scientists whose minds are also embedded in con-crete? However, this is not unusual. Throughout history, each vi-sionary who carried the world ahead had a pack of his colleagues snapping at his heels and trying to drag him back, whether in sub-conscious jealousy or in sheer mental myopia we don't know. A classic case is Dr. Simon Newcomb, who at the turn of the century wrote a mathematical treatise *proving* that the heavier-than-air machine could never carry any load greater than one mosquito (a gram). That was in 1902, one year before the disrespectful deed of the Wright Brothers at Kitty Hawk.

The Simon Newcombs of space are with us today. We will leave them mercifully anonymous in the following authentic quotes of (alleged) experts:

—1946, consensus of a poll among leading scientists: "Space flight will not come before 1975, and probably not before the end of this century."

—1952, a scientist tops in medical research (only): "The ICBM is virtually impossible. We can forget it."

—1956, a famed inventor: "Men will never fly into space and reach the moon."

—1957, August, an IGY official: "America will of course be first in space with a Vanguard satellite. What other nation could beat us?" (Sputnik I was orbited on October 4).

—1957, October 5, Budget Bureau chief: "Sputnik will be forgotten in six months."

—1961, April 9, a space medic training the astronauts: "From studying our Mercury Program, I can state man isn't ready to fly into space for at least another year." (Gagarin went up three days later, our own Shepard the next month—safely.)

This anti-space and/or anti-man-in-space prejudice still exists, unfortunately, among a die-hard group of "black box" scientists who believe machines can do better than men. They refuse to listen when reminded that no space monitor's computer today can foreseeably be programmed to equal a tiny fraction of the human brain's fantastically wide range of perception.

The machine brain must be told *what* to look for, item by weary item. For instance, using a fanciful but forceful example, a computer told to report any *green* dragons flying in space would ignore a flock of *red* ones that flitted all around.

To take a more realistic situation, the Surveyor moon-recon vehicle that is to be soft-landed on the moon by 1964 will, among other things, drill 6 feet down to bring up core-samples of lunar sub-soil. It is ingeniously geared to pause and take a sample every six inches on the way down. It would ingenuously fail to report anything *between* two samplings—which might be a vein of pure radium 3 inches thick.

What two-legged surveyor would miss that, or anything else?

The surprises awaiting mankind on other worlds and all through space are inconceivable today—even to the king of all dreamers. Perhaps one scientist has summed up best for you what the Space Technology "Race" (not against Russia but simply from gathering momentum of its own) really means:

"Before us lies the Age of Applied Imagination. In checking which group has been most consistently close in foretelling some of the space discoveries and events that have already come about, I find science-fiction writers way out in front for the past 25 years. If we want to get ahead fast, let us not waste time but refer directly to Buck Rogers . . . as the starting point."

The writer of this book originally produced science-fiction. Again, to merely report facts, in a 1953 poll among 65 space experts, only eight of us predicted the first unmanned lunar vehicle before 1960 (Lunik I and Pioneer IV made moon flybys in 1959). Of those eight, five were science-fiction writers. The majority of the other 57 men, who chose anywhere from 1970 to the next century, included every professional scientist in the poll but one (Dr. Fred Whipple, Director, Smithsonian Astrophysical Observatory, Cambridge, Mass.)

One might draw from this the apparent conclusion that the over-cautious traditions of science tend to smother any truly realistic—which they consider "optimistic" or even "wild"—projections into future scientific progress. This conservatism naturally filters from the authorities to the people, old and young.

If you, for instance, had sat down in 1957, just prior to Sputnik, would you have predicted all that has come true as of today:

—100-plus USA space vehicles launched, almost 50 of them still in orbit?

—A 100-foot "balloon" circling earth for two years, visible to all the people on earth?

—A probe circling the moon and photographing its hidden side?

—A probe reporting Venus mysteries during flyby?

—Manned orbital flights becoming almost routine, for durations of days?

—Discovery of Van Allen Belts, solar "death-ray" storms, a ring of dust and hydrogen-tail for earth?

And would you have dared make these predictions involving our national budget and economy:

—An Apollo moonship budget of $20 billion at a minimum, passed by jaundice-eyed Congress?

—Such great companies as Ford, GM, GE, Convair, Lockheed, Boeing, all jumping on the space bandwagon and earning from 25% to 90% of their gross income from spacecraft rather than earthcraft?

—200,000 small business firms linked into a vast supply-chain for the enormously expanded astronautics industry?

—25% of the nation's pool of physical scientists and non-agricultural engineers laboring day and night on missiles and rocket craft?

Do you think that—back in August of 1957, just before an eerie "beep-beep" echoed from space—you would have foretold the President of the United States firmly declaring that America will be in the forefront in "sailing the ocean of space," and thereby launching a national policy committing 190 million Americans to exploration of the entire solar system?

In fact, no one on earth—*no one*—dreamed half as wildly about all those space events which came true.

If you were to sit down *now* and write a *future* list of what might come about by 1975, during your space career, do you think you might conceive *one-tenth* of the actual overwhelming picture? Try it. Check in 1975. It might be a valuable guidepost for what to expect after *that* and thereby place you in the forefront in space prophecy and leadership.

Now is the time for you to think in the future. That is where you will spend your lifetime. You will be hopelessly left behind if you use present-day Simon Newcomb thinking. Any resemblance between the world of today and the world of 1975 will be purely impossible. Throw in all your mental horsepower and shift your imagination into high gear. Only then will you really belong in the Age of Advancement.

Good luck to your career in space.

APPENDICES
AND
INDEX

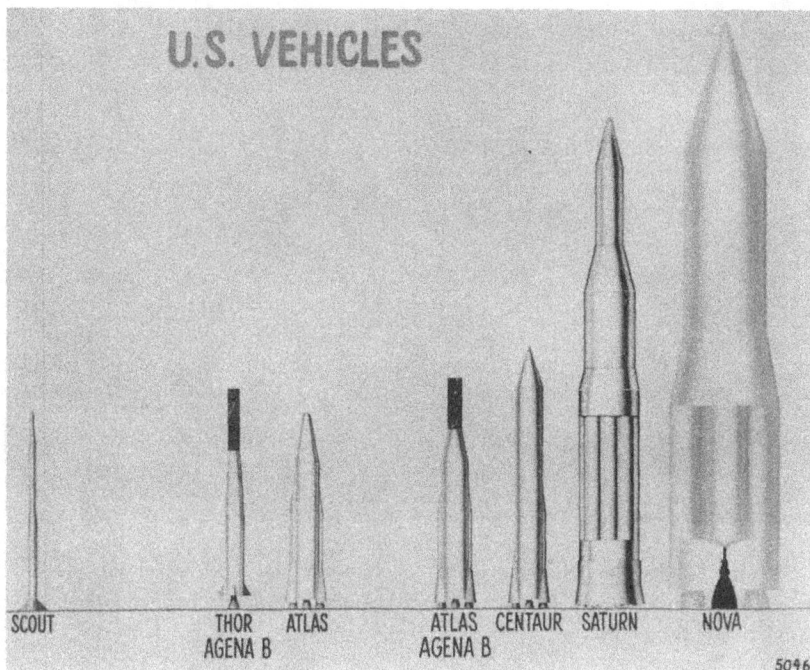

U.S. VEHICLES

SCOUT   THOR   ATLAS   ATLAS   CENTAUR SATURN   NOVA
         AGENA B   AGENA B

5096

These rockets will be America's principal launching vehicles for the next decade. Nova, when it becomes operational, will have a thrust of something in the vicinity of 12 million pounds.

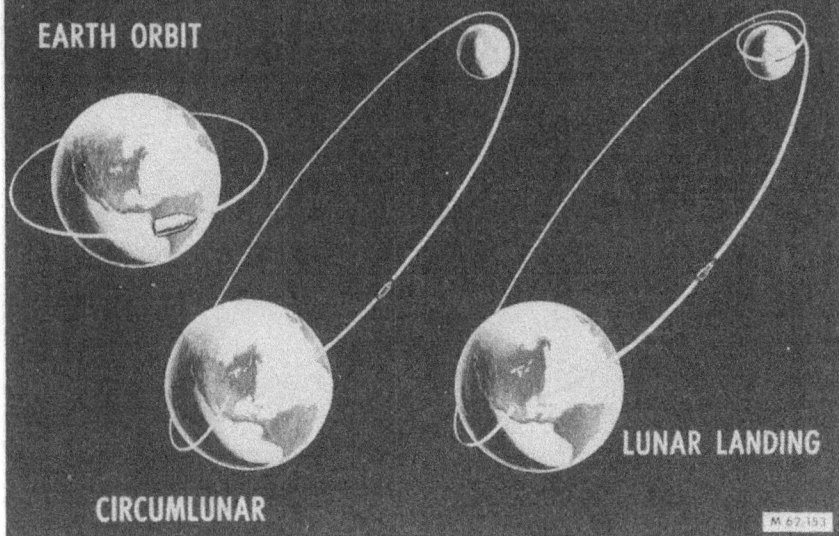

PROJECT APOLLO
THREE MISSIONS

EARTH ORBIT

LUNAR LANDING

CIRCUMLUNAR

M 62-153

Project Apollo, the "moon project," is currently the most important space-exploration venture in America. This NASA diagram illustrates the three-man phases of the project.

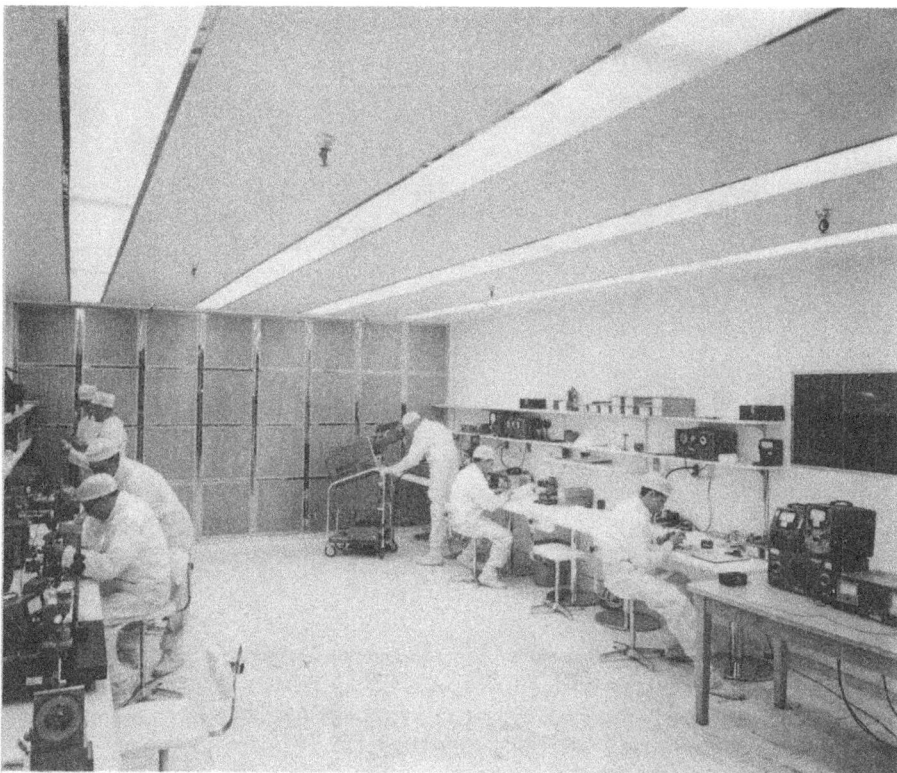

A system of ultra-filtered constantly circulating air makes this room in the General Dynamics/Pomona plant the cleanest in the world. Here technicians in surgical dress assemble sensitive missile parts in an area 200 times cleaner than a hospital operating room.

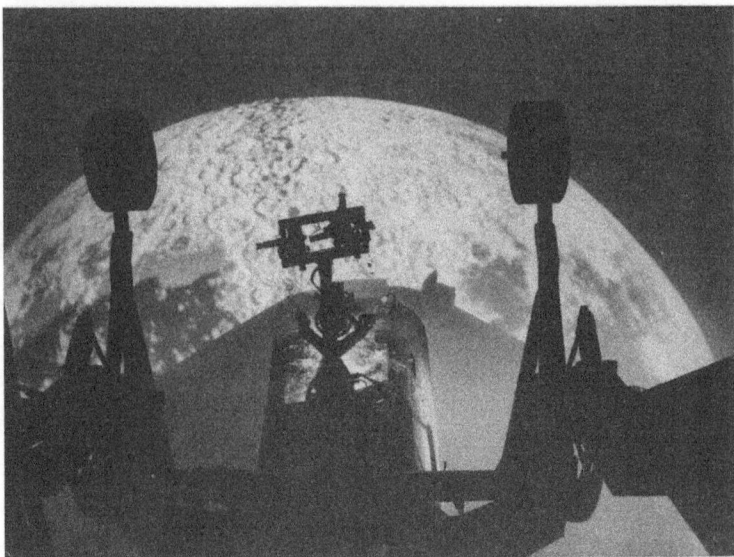

Ling-Temco-Vought has developed this amazing "Manned Space Flight Simulator." Coupled to a room full of computers, the device permits the astronaut to experience many facets of space flight, space rendezvous and lunar landing without leaving the ground.

Julian Palmore III, while at Cornell University, designed an unmanned lunar landing vehicle. The design, for which Palmore won the $1000 American Rocket Society-Chrysler Corporation Award, was used in the construction of the lunar Ranger vehicle.

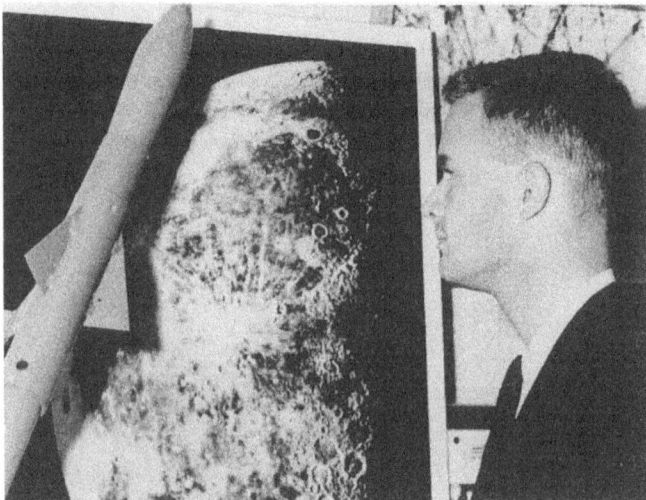

Missile and space vehicle components weighing up to 300 pounds are subjected to 100 g's of force on a 36-foot-diameter centrifuge in the Douglas Aircraft Company's acceleration laboratory.

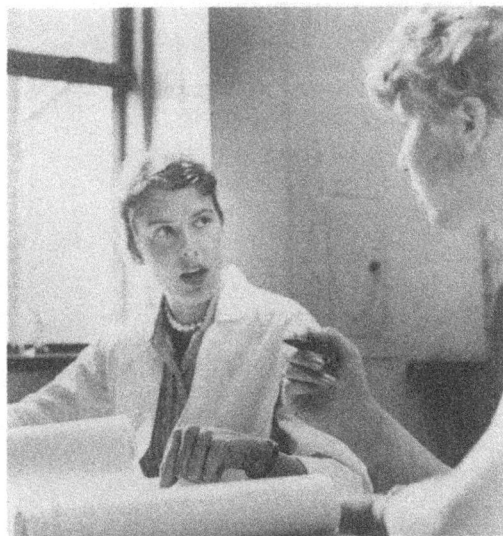

Biologist Peggy Hazard *(left)*, a recent graduate of Bucknell University, is now working on one of Bell Telephone Laboratories' space projects. Here she discusses a problem with chemist Moll Sibul, another of the more than 2500 women employed by Bell.

APPENDIX 1

Educational Requirements and Opportunities for Careers in Space

TABLE A   Descriptions of the most important specialities in Aero-Space Technology   Source: NASA

Described below are the major occupational fields of Aero-Space Technology and the specialties within each field. The first specialty in each field is of a broad general nature including aspects of several or all of the other specialties in that field.

See page 18 for table of specialties and locations.

These specialties represent NASA work as of the date of this announcement, but due to the rapidly changing frontiers of aero-space technology they are subject to change without published amendment. When a new specialty is recognized, applications already on file will be reviewed, and candidates who are found to be eligible for the new specialty will be rerated for the additional specialty.

FLUID AND FLIGHT MECHANICS

Fluid and Flight Mechanics.—Aerodynamic and space vehicle studies and vehicle environment studies of a general or varied nature, including aerodynamically and mechanically induced loads, vibrations, aeroelasticity, noise, internal flows, fluid flows, boundary layers, heat transfer, skin friction, controls, and other areas of fluid and flight mechanics within and beyond the earth's effective atmosphere.

Flight Mechanics.—Responses of vehicles to forces, flows, and associated phenomena encountered in flight within and beyond the earth's atmosphere; flight path calculations including trajectories, ranges, orbits, reentry aerodynamics, and forces and moments affecting bodies in flight; aerodynamic, inertial, gyroscopic, and other effects on oscillatory and dynamic characteristics; configuration and matching of flight vehicles and launch vehicles; space rendezvous; energy management studies related to flight missions.

Fluid Dynamics.—Flow fields around bodies in terrestrial and other atmospheres at all speed ranges through hypersonic; correlation of transitional points and types of pattern from laminar to turbulent flow; development of experimental techniques for testing and photographing; studies of variables in transitional patterns; pressure distributions; separated flows; vorticity studies, velocity distributions; transition and breakdown of local flows.

Magnetofluiddynamics.—State variables, transport properties, radiative behavior, power requirements, and dynamic characteristics of ionized gases, plasmas, and other conducting fluids under the action of magnetic and electric fields; electromagnetic pumping phenomena; applications to space flight and reentry problems.

Aerostructural Dynamics.—Free, forced, and self-induced vibration of structures of aircraft, missiles, space vehicles, and launch equipment; inertial, aerodynamic, and structural parameters; flutter, divergence, aeroelasticity, structural feedback; atmospheric turbulence; acoustic fields; dynamics of inflatable and erectable space structures; landing impact of aircraft, returning space vehicles, and interplanetary vehicles as examples of forced excitation.

Flight Vehicle Acoustics.—Theoretical and experimental study of effects of noise on design and operation of aircraft, missiles, and space vehicles; fundamental phenomena of noise generation, transmission through air and complex structures; techniques and devices to minimize effects of noise and shock waves on personnel, structures, and sensitive mechanical and electrical equipment; generation and propagation of shock waves; atmospheric and terrain effects; stress response and fatigue characteristics under intense noise loading; devices and operational procedures for minimizing effects of noise.

Propulsion System Dynamics.—Bulk properties of internal and exit air flow, including free-flow and propulsion-induced streams, affecting aerodynamics of aircraft and spacecraft; viscous effects, pressure distributions, thermal effects, airstream calibration, jet exit effects, interaction of jet bulbs or plumes; study of interrelationships of these and related factors to produce optimum geometric and dynamic configurations by integration with airframes.

Heat Transfer.—Aerothermodynamics applied to problems of supersonic and hypersonic flight, reentry, flight in planetary atmospheres; shock-boundary-layer interaction; radiative, conductive, and convective heat transfer; real gas effects, ablation, heat sinks, transpiration cooling; hypervelocity and hyperthermal experimentation.

Stability and Control.—Input, output, and vehicle response characteristics and requirements in launches, flights, and landings of spacecraft and aircraft; gyroscopic effects; non-linear pitching moments; oscillatory motions; investigation and prediction of stabilization.

Control and Guidance Systems.—Integrated systems for control and guidance of vehicles in aerodynamic and space environments including fully automatic and human participation systems; dynamics of vehicles and parameters in control and guidance; design and performance specifications for systems for specific missions; experimental and theoretical studies including analog and other simulation devices.

Trajectory-Orbit Analysis.—Prediction of trajectory requirements for specific missions; premising of hardware design requirements including vehicle, launching, and control systems; interplay of specific mission systems and space mechanics, including consideration of location, motions, and gravitational fields of celestial bodies; orbit determination and prediction; interpretation of flight path perturbations in terms of physical phenomena.

Appropriate College Majors

Engineering Physics, Engineering Mechanics, Astronautics, Aeronautical Engineering, Mechanical Engineering, Physics (except major in nuclear physics), Electrical Engineering (except major in production, transmission, and use of large scale industrial electric power), or other appropriate field of physical science or engineering; or major in Mathematics supplemented by at least 18 semester hours in some combination of appropriate physical science or engineering courses.

Any of the undergraduate majors provided in the Basic Education Requirement is acceptable for applicants with the required graduate study and/or professional experience that is closely related to one of these specialties.

ENERGY AND POWER SYSTEMS

Energy and Power Systems.—Systems and phenomena involved in converting energy into propulsion and/or power for aircraft, launch vehicles, and space vehicles; potential sources of energy; performance parameters for nuclear ramjets, turborockets, other advanced energy mechanisms; environmental testing and component matching; development and application of facilities and testing conditions.

Nuclear Propulsion and Power.—Studies of generation, transfer, and conversion of nuclear energy for propelling or operating aircraft and space vehicles; nuclear systems analysis; nuclear processes, shielding, reactor core physics, operational problems.

Electrical Propulsion and Power.—Studies and applications of magnetofluiddynamics and plasma physics; steady and unsteady electrical and magnetic field generators; electron and ion sources, surface ionization, ion-electron mixing processes; plasma sources and accelerators; photo-voltaic, thermal-electric, and related sources for internal spacecraft power supply; environmental effects including solar radiation, outgassing and leakage, and heat transfer.

Chemical Propulsion and Power.—Chemical conversion systems for aircraft and space-vehicle propulsion and power; rockets, jets and related systems; cryogenic propellants; batteries, fuel cells, and related sources for internal spacecraft power supply; synthesis of new liquid and solid propellants; combustion stability; thermal conductivity and viscosity; component cooling problems.

Internal Flow Dynamics.—Fundamental aerodynamics, hydrodynamics, and thermodynamics applied to energy and power systems, fueling, and other aeronautical and space-flight systems and components including pump lines, valves, turbomachinery, tank flows, cryogenic propellants, cavitation, boundary layer separation; dynamics of fluid masses and flows under weightlessness.

Friction and Lubrication.—Support, lubrication, and sealing of rotating and sliding mechanisms peculiar to aero-space flight and propulsion systems; environmental effects including high-temperature, cryogenic, chemical, nuclear, and vacuum conditions; additives, inhibitors, and protective coatings; lubricant fluids and materials; physical mechanics of friction.

Appropriate College Majors

Chemical Engineering, Chemistry, Nuclear Engineering, Electrical Engineering (if included at least 1 course in thermodynamics or nuclear physics), Engineering Physics, Engineering Mechanics, Physics (classical or nuclear), Aeronautical Engineering, Mechanical Engineering, Astronautics, or other appropriate field of physical science or engineering; or major in Mathematics supplemented by at least 18 semester hours in some combination of physics, thermodynamics, chemistry, or closely related fields.

Any of the undergraduate majors provided in the Basic Education Requirement is acceptable for applicants with the required graduate study and/or professional experience that is closely related to one of these specialties.

MATERIALS AND STRUCTURES

Materials and Structures.—Dynamic loads, high and low temperatures, particle impact, radiation, erosion, and atmospheric and space environmental effects on structures and materials of spacecraft, aircraft, launch vehicles, and launching systems.

Structural Mechanics.—Behavior of structural elements and assemblies under environmental conditions including elevated temperatures and interactions with environments; determination of such structural responses as stresses, deflections, and temperature; determination of vibration, fatigue, and creep characteristics; steady state loads; determination of dynamic and static stability.

Flight Structures.—Investigation, development, and evaluation of new and improved concepts for structural design of aircraft, spacecraft, launch vehicles, and launching systems; strength and weight analysis methods; design criteria for structural elements and assemblies; structural concepts and parameters in design and performance of aero-space propulsion systems.

Structural Materials.—Strength of materials for aero-space vehicle structures; fatigue; strength loss at high temperatures; oxidation and ignition of metals in high temperature; melting, fusing, and recrystallization techniques and effects; ablation and mass-transfer cooling; chemical cooling; emissivity of materials; erosion and sputtering; corrosion; protective coatings; adhesives; high polymers and plastics; related aspects under high-speed flight, reentry, and extended spaceflight conditions.

Basic Properties of Materials.—Solid state mechanisms of materials for atmospheric and space flights; atomic and molecular structures; cryogenic and ultra-temperature effects; structural defects in crystals; diffusion in metals and ceramics; semiconductors; reaction of solids with disassociated gases; chemical-physical interrelationships.

Appropriate College Majors

Ceramics or Ceramic Engineering (if 12 semester hours in refractory ceramics, cermets, protective coatings), Metallurgy or Metallurgical Engineering (if 12 semester hours of physical or adaptive metallurgy, high-temperature metals and alloys, cermets), Physics, Engineering Physics, Aeronautical engineering, Mechanical Engineering, Civil Engineering (if 12 semester hours of strength of materials, structures, thermodynamics), Chemistry, or other appropriate field of physical science or engineering; or major in Mathematics supplemented by at least 18 semester hours in some combination of physics, structures, materials, thermodynamics, or other appropriate courses. In addition for *Basic Properties of Materials:* at least one course in atomic or nuclear physics.

Any of the undergraduate majors provided in the Basic Education Requirement is acceptable for applicants with the required graduate study and/or professional experience that is closely related to one of these specialties.

SPACE SCIENCES

Space Sciences.—Forces, motions. and states in space; gravitational, magnetic. and electric fields in space; energetic particles and particle populations in space; solar-terrestrial and solar-planetary relations; interplanetary plasma; celestial mechanics and geodesy; solar and planetary atmospheres, ionospheres, structures, and developmental history; cosmology and relativity; related studies, both theoretical and experimental.

Aeronomy.—Earth and planetary atmospheres including composition, structure, and motion of gases; interactions of incident radiation, particles, and aggregate matter with such atmospheres.

Ionospheric Studies.—Earth and planetary ionospheres and regions of transition between the ionospheres and the interplanetary plasma.

Particles and Fields.—Particles of energy greater than thermal, such as cosmic rays, radiation belts of earth and planets, auroras; magnetic and gravitational fields of the sun, planets, and space.

Stellar and Galactic Studies.—Stellar and galactic astronomy, including origins and nature of the universe; the sun and its emanations.

Lunar and Planetary Studies.—Structure, composition, and immediate environs of moons, planets, asteroids, and comparable bodies in the solar system; interplanetary medium and its dynamics; developmental history of planets and their moons.

Appropriate College Majors

Physics, Astronomy, Meteorology, Geology, Geophysics, Astro-physics, or other appropriate field of basic physical science applicable to a specialty of AST Space Sciences, including or supplemented by at least one physics or engineering laboratory course in electronics, optics, materials, vibration, high-vacuum theory, heat transfer, or comparable field relating to aero-space instrumentation.

Any of the undergraduate majors provided in the Basic Education Requirement is acceptable for applicants with the required graduate study and/or professional experience that is closely related to one of these specialties.

At all grade levels, applicants for positions in Space Sciences specialties must have had a background of education and/or experience which will enable them not only to perform the space sciences aspects of the work, but also to understand, use, and interpret the highly specialized ground-based and in-flight measurement and control equipment required for space science observations and experiments.

MEASUREMENT AND CONTROL SYSTEMS

Measurement and Control Systems.—Planning and integration of measurement and control systems required for aero-space research in flight vehicles, launch vehicles, wind tunnels, free flight facilities, shock tubes, flight simulators, and other research vehicles and facilities where the systems require specialized skills and knowledges peculiar to or characteristic of aero-space programs; design and development of complete integrated systems including electrical, electronic, mechanical, optical, and electromechanical components for sensing, measuring, transporting, recording data, or controlling such factors as temperature, forces, pressures, loads, velocities, accelerations, altitudes, density levels, flow patterns, shock waves, ionization levels and rates, dissociation levels and rates, proton densities, and energy levels.

Sensors and Transducers.—Research, design, and development and sensors for detecting and measuring phenomena and transducers for translating such measurements into other forms for transmission or recording, requiring specialized skills and knowledges peculiar to or characteristic of aerospace programs; measurement of physical, electrical, chemical, color, or other phenomena associated with such variables as displacements, inertia, flow of electricity, gas excitations, pressure differentials; development of such types of sensors and transducers as pressure cell diaphragms, quartz crystal transducers, gyroscopes, strain gages, photoelectric cells, thermocouples, geiger counters, bolometers, inertia sensors, and light path sensors for use in plasma experiments, bio-packs, cryogenic research, and a wide variety of other experimental and environmental conditions.

Heat and Light Measurement.—Research, design, and development of sensing and measuring techniques and devices requiring specialized aero-space skills and knowledges, for thermal, light, and associated excitations and emanations; development and use of thermocouples, thermisters, thermopyles, pyrometry, spectrography, and other instruments and techniques.

Radio Frequency Systems.—Planning and development of components and complete aero-space systems for signal generation, modulation, transmission, reception, demodulation, and recording; telemetry, radar, ionization probes, video-frequency, audio-frequency, tracking, command-control, and related systems for aero-space programs; use of masers, tracking filters, magnetic recorders, parametric amplifiers, antennas, and related equipment.

Force Measurement.—Design, development, evaluation, and application of techniques and devices for measurement of forces, including strain gage balances with extreme force and temperature ranges and static, transient, and fluctuating force measurement problems; application to flight vehicles, launch vehicles, space structures, and a variety of experimental facilities and equipment requiring specialized skills and knowledges peculiar to or characteristic of aero-space programs.

Measurement Standards and Calibration.—State-of-art studies of equipment, systems, and methods used in aero-space environmental testing and calibration of electronic, electrical, mechanical, and liquid-flow devices; establishment of entirely new basic standards for measurement, environmental testing, and calibration of novel and unique aerospace research sensing and measurement devices; performance and reliability of such devices under conditions peculiar to or characteristic of aerospace environments.

Automated Control Systems.—Mission-oriented requirements for flightpath accuracy, vehicle stability, and other aspects of aero-space vehicles; dynamics and kinematics of spacecraft and aircraft; research and development of electronic, mechanical, electromechanical, and inertial control, guidance, stabilization, feedback, servo, and navigation systems and components; transient and dynamic behavior of systems and components; sequencing and programming of complete integrated systems; effectiveness and reliability studies.

Appropriate College Majors

Electronics, Electronic Engineering, Electrical Engineering (except major in production and use of large scale industrial power), Physics, Engineering Physics, Mechanical Engineering, or other appropriate field of physical science or engineering. Such major should have included at least two of the following courses: solid state physics, materials, optics, statics and dynamics, electricity and electronics, electron optics, kinetic theory of gases, heat transfer, electromagnetic propagation or radiation, semiconductors, vibration, high-vacuum theory, information theory.

Any of the undergraduate majors provided in the Basic Education Requirement is acceptable for applicants with the required graduate study and/or professional experience that is closely related to one of these specialties.

DATA SYSTEMS

Data Systems.—Application of mathematical and computer theory and methods to data reduction, computation, and simulation problems of experimental research with experimental aircraft, space flight vehicles, launch vehicles, and ground-based aero-space research facilities; participation in design of experiments, evaluation of mathematical assumptions, range and calibration of sensing and recording instrumentation, and compatibility and integration with computing and read-out systems; evaluation, procurement, modification, application, and operation of computing equipment.

Theoretical Computation Techniques.—Application of digital computer theory and methods to the formulation and study of mathematical expressions of aero-space research and development problems, requiring advanced comprehension of physical aspects of problems and close collaboration on aero-space programs; formulation analysis of the most feasible basic mathematical approaches and efficient methods of solving and computing aero-space problems.

Theoretical Simulation Techniques.—Application of analog computer theory and methods to the formulation and study of theoretical models of aero-space research and development problems, requiring advanced comprehension of such fields as flight mechanics, orbit-trajectory analysis, control and guidance; close collaboration on aero-space programs; adaptation and modification of analog computers and other simulator facilities; development of new simulation techniques.

Appropriate College Majors

Either (A) a major in mathematics supplemented by at least 12 semester hours in some combination of appropriate physical science or engineering courses, or (B) a major in physical science or engineering including or supplemented by at least 6 semester hours of mathematics beyond the calculus. Included or in addition, applicants must have had at least two courses in subjects closely related to data systems techniques (such as computing systems, digital logic, analog systems) or data systems equipment (such as electronics, electrical networks).

Any of the undergraduate majors provided in the Basic Education Requirement is acceptable for applicants with the required graduate study and/or professional experience that is closely related to one of these specialties.

At all grade levels, applicants for positions in Data Systems specialties must have had a background of education and/or professional experience which will enable them not only to perform the data systems aspects of the work, but also to assimilate readily the physical aspects of problems in aero-space research, development, and operations and to collaborate closely with other professionals in the design of experiments and the interpretation of data.

EXPERIMENTAL FACILITIES AND TECHNIQUES

Experimental Facilities and Techniques.—Analysis of research requirements for facilities, models, vehicles, and equipment for conducting experimental research on aero-space problems; development of specifications or designs for prototype or unique experimental facilities and equipment; evaluation of contractors' proposals; supervision and coordination of work of engineering design groups to evolve reliable, economical, and versatile facilities and equipment.

Experimental Facilities.—Development and design of a variety of experimental facilities to meet research requirements in terms of configuration, specialized loadings, dynamic forces, and special sealing and expansible structure requirements; pressure and vacuum vessels of large size or extreme performance limits; planning of specialized facility site locations and collateral requirements to enable world-wide range operations and other experimental activities; launch vehicle static test rigs; vibration, rotation, and other test equipment; surveillance over critical technological aspects of contractors' proposals and operations.

Experimental Equipment.—Development of specialized equipment and models for aero-space experimental programs requiring analysis of program objectives; planning of articulations, special devices, and machinery such as balances, space structural models, arc chambers, remotely controlled models, high-temperature high-load equipment, and other devices; surveillance over critical technological aspects of contractors' proposals and operations.

Fluid and Flow Systems.—Development and design of systems and equipment for achieving, changing, and controlling the physical states of gases, liquids, and solids, including highly toxic, volatile, unstable, and corrosive substances, under conditions peculiar to aero-space experimentation; facilities and techniques for compressing, storage, piping, evacuating, heating, drying, purifying, purging, cooling, metering, cycling, and mixing of aeroform fluids and liquids under extreme conditions or to meet requirements such as cryogenic conditions, extremely high temperatures and pressures, or near absolute vacuums; research and experimentation to develop new design concepts for such systems; evaluation of technological aspects of contractors' proposals, and surveillance over critical technological aspects of design and construction.

Electrical Experimental Equipment.—Development and design of specialized aero-space equipment involving electrical components, devices, and apparatus; electrical power and controls for ion accelerators, fuel heaters, electron grids, vacuum diffusion pumps; magnetic and electrical field apparatus applicable to plasma testing rigs; related research and experimentation; contractor proposal evaluation and surveillance.

Experimental Facility Techniques.—Developmental work for technological operation of aero-space experimental facilities and equipment for research, development, and testing objectives involving highly specialized aero-space knowledges and skills in addition to mechanical, electrical, and other aspects; scheduling, operation, maintenance, calibration, and use-analysis of various types of aero-space facilities including altitude tanks, hypervelocity ballistic ranges, wind tunnels, vacuum chambers, engine test cells, launch vehicle testing rigs, lubrication and power distribution systems, electrical and air dispatching and routing; operational testing, monitoring, and evaluation of facilities; emergency failure planning and action; close coordination with research, development, and facility-development aero-space specialists.

Nuclear Facility Techniques.—Operation, maintenance, and safety of nuclear reactor for experimental research on nuclear materials, components, and shields for specialized aero-space programs; protection and decontamination of machinery and process systems; cooperation with research and development specialists in planning and analyzing results, requiring knowledge and understanding of problems peculiar to or characteristic of aero-space programs.

Flight and Launch Vehicle Systems.—Complete launch and flight vehicle system development and design for aero-space flight programs, involving consideration of thermal effects, center of gravity and center of mass requirements, aerostructural design and stress analysis, engines, tankage, nozzle configuration studies, control and guidance devices, staging and stage separation mechanisms, power sources, payload and instrument packaging, and re-entry and recovery systems; compatibility of ground handling, launching, and tracking facilities; integration with mission requirements and scheduling; evaluation of technological aspects of contractors' proposals, and surveillance over critical technological aspects of design and fabrication activities, both agency and contractor directed.

Launching Systems and Techniques.—Development and use of systems and techniques involved in aero-space launchings, including coordination of flight testing, specialized launching devices and flight procedures, safety techniques, mobile and fixed systems; evaluation of contractors' proposals and surveillance over agency and contractor activities; coordination and cooperation in joint efforts with other governmental and industrial organizations.

Flight Vehicle Experimental Techniques.—On basis of manned systems analysis requirements, planning and development of components, sub-systems, and complete systems involving manned operation and control of experimental flight vehicles and supporting equipment; design of manual flight controls, instrument displays, monitoring and override controls for automatic devices, and related equipment; monitoring and engineering of flight safety of vehicles, sub-systems, and equipment; planning and direction of flight vehicle pre-flight preparation procedures; surveillance over contractor and agency activities; coordination and cooperation in joint efforts with other governmental and industrial organizations.

Appropriate College Majors

Mechanical Engineering, Electrical Engineering, Electronic Engineering, Aeronautical Engineering, Chemical Engineering, Nuclear Engineering, Structural Engineering, Physics, Engineering Physics, Engineering Mechanics, Metallurgy, Ceramics, or other appropriate field of engineering or physical science.

Any of the undergraduate majors provided in the Basic Education Requirement is acceptable for applicants with the required graduate study and/or professional experience that is closely related to one of these specialties.

TABLE B   Curricula available to students wishing to specialize in astronautics   Source: *Careers in Astronautics,* April, 1959

Below is a summary of the technical disciplines which pertain to the field of astronautics. The letters and numbers which appear on this table form the key for the chart which follows.

On the top of the following chart, Topic Areas and their Subdivisions are listed in bold face type. Beneath them, opposite various educational institutions, are listed symbols which apply to the specific Topic Area Subdivisions directly above at the head of the column. The meaning of the symbols is as follows: numbers (e.g. 1, 4, etc.) indicate the total number of courses available in the appropriate Topic Area Subdivision; the letter "x" indicates that one or more courses are offered but that the exact number is not known; the letter "e" indicates that the Subdivision is extensively covered by at least six different courses.

TECHNICAL DISCIPLINES IN ASTRONAUTICS

(Socio-humanities not included)

A. Fundamental Sciences

1. Mathematics: Differential and integral calculus; analytic geometry; partial differentiation; differential equations; Laplace transform; Fourier series; boundary value problems.

2. Classical Physics: Newtonian mechanics; statics and hydrostatics; rigid body dynamics; oscillations and waves; heat and kinetic theory; electricity and magnetism; optics, atomic physics.

3. Basic Chemistry: Atomic structure; mass and energy relationships; rate and equilibrium of chemical reactions; ionic chemistry; acid-base systems; crystals and molecules.

B. Applied Sciences

1. Statics and Dynamics: Forces and moments, stability and instability; kinematics of particles; work and energy; impulse and momentum; rigid-body dynamics; linear oscillatory systems; free and forced vibrations including damping.

2. Solid Mechanics and Physics of Materials: Force transmission; stress-strain relations; stress distribution and deformation; beam theory; shear and torsion; buckling and instability; material structure; material failure; plasticity; stress-strain-temperature-time relations; stress concentration, brittle failure, and fatigue.

3. Electrical Science: Principles of electrical circuits; components and systems for power, control, and instrumentation; theory and performance of electrical components, circuits, and systems.

4. Fluid Mechanics and Gasdynamics: Dynamics and thermodynamics of real and perfect fluids; one-dimensional flow; incompressible and compressible flow, viscous and turbulent flow; lift, drag, and boundary layer effects.

5. Thermodynamics: Thermodynamic laws; properties of liquids, vapors, and gases; heat transmission; gas and vapor cycles.

C. Professional Engineering Subjects

1. Mechanics of Orbits and Trajectories: Celestial mechanics; orbits and perturbations; space flight trajectories; exterior ballistics of rockets; powered flight trajectories; optimization techniques.

2. Propulsion for Rockets, Missiles, and Space Vehicles: Fundamentals of rocket propulsion; solid and liquid chemical rocket engines; rocket components and accessory design and performance; nuclear, plasma, ion, and other systems.

3. Rocket and Space Vehicle Structures and Materials: Analysis and design of shells and pressure vessels; elastic and plastic response to dynamic loads; applied thermoelasticity and aerothermoelasticity; basic loads and materials for missiles and space vehicles.

4. Guidance, Navigation, and Control for Missiles and Spacecraft: Inertial, celestial, and electronic navigation techniques; missile guidance systems; control system design for missile and/or space vehicle.

5. *Vehicle Design:* Rocket vehicle performance; missile and/or space vehicle design; practice in system design combining propulsion, mechanics, aerodynamics, structures, guidance, and control for missiles, re-entry vehicle, space vehicle, etc.

D.   Professional Sciences

1. *Communications:* Radar systems; ultra-high frequency techniques; information theory; microwave circuits; electromagnetic radiation theory and antenna design; telemetering; pulse circuit analysis.

2. *Astronomy:* Planetology; study of the sun, asteroids, comets, and meteors; astrophysics; physical astronomy.

3. *Physics of the Upper Atmosphere:* Composition, properties, meteorology of the upper atmosphere; tools of upper atmosphere research.

4. *Hypersonic and Rarefied-Gas Dynamics:* Superaerodynamics; free molecule flow; slip flow; piston theory; boundary layer and heat transfer effects for rarefied gases and gases at high temperature; plasma physics; magnetohydrodynamics.

5. *Space Medicine:* Physiological and psychological aspects of environment of upper atmosphere and space; artificial environments and environmental control.

E.   Related Advanced Science and Engineering

1. *Advanced Mathematics:* Probability and operations analysis; statistical theory; advanced calculus; variational calculus; field theory; machine analysis and digital techniques.

2. *Aeronautical Sciences:* Three-dimensional flow; wing theory; classical performance, stability, and control; aircraft propulsion; aeroelasticity.

3. *Modern Physics, Solid State Physics:* Quantum theory of matter; quantum mechanics; theory of molecular structure; interaction of matter with electric and magnetic fields; insulators, semiconductors, metals, and molecular compounds; relativistic mechanics and dynamics.

4. *Nuclear Technology:* Nuclear structure; nuclear reactions; reactor theory and design; shielding; reactor control.

5. *Physical Chemistry:* Composition and physical states of matter; chemical thermodynamics; kinetic theory of gases; properties of solutions; kinetics of chemical reactions; surface and colloid chemistry; atomic and molecular spectra; electronic structure of molecules; theory of combustion; transport phenomena.

6. *Meteorology and Geophysics:* Physical meteorology; planetary atmospheric dynamics; geochemistry; geophysics; geodynamics; geodesy.

7. *Metallurgy and Materials:* Metallurgical science; advanced materials science; physics of strength and plasticity.

8. *Automatic Control and Servomechanism Theory:* Linear representations of physical systems and analysis of their performance; stability criteria, transient response, frequency response; feedback loops; performance functions by Laplace transforms; analytical, graphical, and analog methods; advanced servosystem analysis and design.

9. *Structures:* Advanced mathematical theory of elasticity and plasticity; advanced shell theory, theory of shallow shells; large deflection theory; methods of optimum design; experimental stress analysis.

DETAILED DATA ON TECHNICAL CURRICULA IN FIELD OF ASTRONAUTICS

	Topic Areas in Which Courses Are Presented			Opportunity for Research Work
	C 12345	D 12345	E 123456789	
Alabama				
Ala. Poly. Inst.	xxxx	xxx	xxxxxxxx	C3, 4; D1; E1, 2, 3, 4, 5, 8, 9
Univ. of Alabama	1x x	111x	eee1x1e1x	C3; E1, 2, 3, 7, 9
Arizona				
Univ. of Arizona	42e22	eex42	eeee4ee4e	C2, 3; D1; E2, 4, 9
California				
California Inst. of Tech.	3e21	4e 2	e3e3e ee3	C, D, E extensive
California State Poly. Tech.		x	xx　x	
Northrop Aero. Inst.	114	1	1123	C3, 4; D1, 4; E (all)
Stanford Univ.	1151	e 15	ee3e1e3e	C, D, E extensive
Univ. of California	3233	eee	eeeeeeee	C1, 2, 3, 4; D1, 4; E1, 5, 6, 7, 8, 9
Univ. of Cal. at Los Angeles	42xx1	33122	exx221 ee	none
Univ. of Santa Clara	xx	x x	x x x x	C2, 3, 4; D1, 4, 5; E1, 3, 5, 7, 8, 9
Univ. of So. California	13ee1	exxex	eexe xee	
Colorado				
Colorado State Univ.	1		2 121 212	C2, 3, 4, 5; D1, 2, 3, 4, 5; E1, 2, 3, 5, 6, 7, 8, 9
Univ. of Colorado	13312	1e1	2ee12e221	D1, 3; E1, 5, 7, 8
Univ. of Denver	1	21	4 324 432	
USAF Academy	x1 1x	1 x1	11 11	
Connecticut				
Yale Univ.	12ex	ee 1	e2e2e eee	C, D, E
Univ. of Bridgeport	1	1	21 12	C, D, E
Univ. of Connecticut	11	x	x1 x x1x	E1, 5
Trinity College	2		4 5 e	
Delaware				
Univ. of Delaware	x	x x	x x2x 13x	C2, 3; E1, 3, 4, 5, 7, 8
Florida				
Univ. of Florida	xxxxx	xxxxx	xxxxxxxx	C2, 3, 4; D1, 4, 5; E2, 3, 4, 5, 8, 9
Univ. of Miami	x	x	xxx x x	None
Georgia				
Georgia Inst. of Tech.	312	e 1	eee4e11 1	C1, 3; D1, 4; E1, 2, 3, 4, 5

Institution				
Idaho				
Univ. of Idaho	111	1111	x2xxxxxx	C2; D1; E1, 4, 5
Illinois				
Bradley Univ.	2 5 1	31	2 222 2	D1; E1, 3, 5, 9
Illinois Inst. of Tech.	13461	3	64916 115	C, D, E extensive
Northwestern		76 4	e6576ee7	C, D, E extensive
Univ. of Illinois	33e11	ee 31	eeee2e2e	
Indiana				
Notre Dame	1	1	exx2e exx	C, D, E
Purdue Univ.	26e21	e2 4	eeeee23ee	C2, 3; D1, 4; E1, 2, 3, 4, 5, 7, 8, 9
Iowa				
Iowa State College	33213	exxxx	eeeeexeee	C, D, E extensive
Univ. of Iowa	1 2	4ex	ee235x 1e	
Kansas				
Kansas State College	111	3 11	ee ee 23	C2; E2, 3, 4, 5, 8
Univ. of Kansas	xxx11	xx 1	eeexexeee	C, D, E
Univ. of Wichita		x	xxx x xxx	
Kentucky				
Univ. of Kentucky		e22	e2eee e5	D1, 3, 5; E3, 4, 5, 6, 7, 8
Louisiana				
Louisiana Poly. Inst.	1 1	11	2	None
Louisiana State Univ.		31	323242222	C3; D1, 2; E1, 3, 5, 8, 9
Southwestern Louisiana Inst.	21 1	2 2	611122231	None
Tulane Univ.		11 1	1 111 11	
Maine				
Univ. of Maine		x	x xxx xx	Yes (unspecified)
Maryland				
Johns Hopkins Univ.	e 1x1 1	x 1x	eeexexxxe	C, D, E extensive
Univ. of Maryland		x113	xexxxlxxx	C1; D2, 3, 4; E1, 2, 3
Massachusetts				
Harvard Univ.	xx	xexxx	exelexexx	C1; D1, 2, 3, 4; E all
Lowell Tech. Inst.	2 2	3	e e56 x3x	D1, 3; E3, 4, 5
Mass. Inst. of Tech.	32253	e126	eeeeeeee	C, D, E extensive
Univ. of Massachusetts	11	21	112241212	D1; E5, 9
Worcester Poly. Inst.	11	2	93434 536	E

(CHART CONTINUED ON NEXT PAGE)

	Topic Areas in Which Courses Are Presented			Opportunity for Research Work
	C 12345	D 12345	E 123456789	
Michigan				
Michigan State Univ.	1121	711 1	717181633	D1; E1, 8, 9
Univ. of Detroit	1 1	1	1111 1	None
Univ. of Michigan	x1x1x	11xx	xexxxxxxe	C, D, E extensive
Minnesota				
Univ. of Minnesota	22e21	e3431	eee3e233e	C, D, E extensive
Mississippi				
Mississippi State Univ.		2	5 1 3 3	D1; E8
Univ. of Mississippi		11	4 224 112	E1, 9
Missouri				
Missouri School of Mines	x		x eeeexe	C3; D1; E3, 4, 5, 6, 7, 8, 9
St. Louis Univ.	1x x	x 2	3xxxxe 4x	D3; E3, 4, 6, 8
Washington Univ. of St. Louis		x	e x x xex	
Montana				
Montana State College	3	e1	4 32e 51	D1; E5, 8
New Hampshire				
Univ. of New Hampshire		41	e eee3322	D1, E
New Jersey				
Princeton Univ.	xx	x	xxx x x	C2; D4; E2
Rutgers Univ.		x	x xxx xxx	
Stevens Inst. of Tech.	xx x	e x	eeexx exx	Yes (not specified)
New Mexico				
New Mexico Coll. of A. and M.A.	11	2 1	6111113	Some
Univ. of New Mexico	x x	xxx	xxxxxxxxx	D, E
New York				
City Coll. of New York	x 24	e22x	e24e645e	None
Clarkson Coll. of Tech.	x	x	x xxx xxx	Yes (not specified)
Cornell Univ.	214 1	ee31	ee56e e5e	Yes
Manhattan College	x	x xx	x xxx x x	
New York Univ.	134	e - 2	e2443ee22	C, D, E extensive
Polytechnic Inst. of Brooklyn	145e1	e 15	e7e6e ee7	C, D, E extensive
Rensselaer Poly. Inst.	1xx1	xx1x1	xexexxexx	C2; D3, 4; E2, 3, 4, 5, 7, 9
Syracuse Univ.	x	x	xxxxxxxxx	C3; D1; E2, 3, 4, 5, 7, 8, 9
Univ. of Buffalo	xx	xx x	xxxxx xx	C3; D1, 4; E1, 2, 3, 8, 9
Univ. of Rochester	xx	55xxx	e ex4xxex	C3; D1, 5; E1, 3, 4, 5, 7, 8
Union College	xx	xx	6x32xxxxx	Yes, not specified

North Carolina				
North Carolina State Coll.	12112	5111	68977 687	C1; D1; E1, 2, 3, 4, 5, 7, 8, 9
North Dakota				
Univ. of North Dakota		x	xxxxxxx x	D1; E1, 3, 4, 5, 9
Ohio				
AF Inst. of Tech.	xxxxx	x xxx	xexxx xex	C, D, E extensive
Case Inst. of Tech.	3221	6e 1	eee6e ee5	C, D, E extensive
Fenn College		1	1 111 1	Yes (not specified)
Ohio Northern Univ.		xx	x xxx x	None
Ohio Univ.		33	2 23e 11	
Ohio State Univ.	35ex1	ee14x	ee7eeeee	C, D, E extensive
Univ. of Akron	x	11	1 11 11	No
Univ. of Cincinnati	xxx x	xx x	exxxxxxxx	C1, 2, 5; D2, 4, 5; E2, 4, 9
Univ. of Toledo	xx	xxx	xxxxxxxxx	Increasing
Western Reserve Univ.		3	1 444 1	Yes
Oklahoma				
Oklahoma State Univ.	xx	x x	xxxxx xx	C2, 3; D4; E1, 2, 3, 4, 5, 8, 9
Univ. of Oklahoma	11	x x	e e e1	
Oregon				
Oregon State College	xx	x x	xxxxxxxxx	Some
Pennsylvania				
Carnegie Inst. of Tech.	1 2	3 1	e1e1x e23	C1, 3; D1, 4; E1, 2, 3, 5, 7, 8, 9
Drexel Inst. of Tech.	121	e	e43ee 3e2	D1; E3, 4, 5
Lehigh Univ.		4 1	e 443 26	C3; D1; E3, 4, 5, 7, 8, 9
Penn. Military College	1	1	1	
Penn. State Univ.	74e42	5432x	eeeee5e	C, D, E extensive
Univ. of Pennsylvania	xxx	ee xx	e3e4e eee	C2,3; D1, 2, 4, 5; E extensive
Univ. of Pittsburgh	xx	x	xxxxx xxx	C3; D3; E1, 2, 3, 4, 9
Villanova Univ.	xx	xx x	e122x 22x	C; E
Rhode Island				
Brown Univ.	1x	93 2	97314 429	C2, 3; D1, 2, 4; E1, 2, 3, 7, 8, 9
Univ. of Rhode Island	1	21	52212 111	D3; E6,7
South Carolina				
Clemson College	1	41	e2e2e 321	E
South Dakota				
South Dakota State College		x1	xxx 21	Yes (not specified)

(CHART CONTINUED ON NEXT PAGE)

	Topic Areas in Which Courses Are Presented			Opportunity for Research Work
	C 12345	D 12345	E 123456789	
Texas				
A. and M. Coll. of Texas	3 x	xxx	4344eex31	D1, 2, 3; E1, 2, 3, 4, 5, 8
Lamar State Coll. of Tech.	x17		x1x1x 12	Limited
Rice Institute	111	1 1	e2e2x xxx	C2, 3; D4; E1, 2, 3, 4
Southern Methodist Univ.	1211	2 1	2e 722 57	D4; E2, 4, 8, 9
Univ. of Houston	1	4	e1414 112	C3; D1; E1, 3, 5, 8
Univ. of Texas	2122	3411	eeeeee2e	C,D,E,
Utah				
Utah State Univ.			3458	Ct
Vermont				
Univ. of Vermont	x	x x	x xxx xxx	D1; E5, 7
Virginia				
Virginia Poly. Inst.	2x	62 2	eeeee3e	D4; E2, 3, 4, 9
Washington				
Seattle Univ.	x	x	xxxxx xx	C3; E5, 7, 8, 9
State Coll. of Washington	xx	xxx	1 111 11	Some
Univ. of Washington	11x	e321	eee5eee5e	C, D, E extensive
West Virginia				
West Virginia Univ.	xx	xxx	xxxxxxxx	C3; D1; E extensive
Wisconsin				
Marquette Univ.	2	xx x	exe4e1e3x	D1, 5; E1, 3, 4, 8, 9
Univ. of Wisconsin	1xxx	exxx	exeeeexx	C2, 3; D, E extensive
Washington, D.C.				
Catholic Univ. of America	xx 21	1	xxxxx x	Yes (unspecified)
George Washington Univ.	1	4	5 2 3 22	D5, 8, 9
CANADA				
British Columbia				
Univ. of British Columbia		x	xxxxxxxxx	D1; E3, 5, 6, 7, 8
Quebec				
Ecole Polytechnique, Montreal	1	1	1 1121x11	C3; D5, 7, 8, 9
New Brunswick				
Univ. of New Brunswick		4	31111 121	E1, 5, 8, 9

TABLE C   Academic Institutions offering Astronautics Study Programs and Fellowships

Source: The American Rocket Society

Educational Institutions	Partial Programs in Astronautics	Programs in Physics	Fellowships in Physics	Programs in Mathematics	Fellowships in Mathematics	Programs in Engineering	Fellowships in Engineering
Ala. Poly. Inst.	*	*		*	7	*	12
Univ. of Alabama	*	*	5			*	3
Univ. of Arizona	*	*	5	*		*	32
California Inst. of Tech.	*	*	60	*		*	103
California State Poly. Tech.	*						
Northrop Aero. Inst., Calif.	*						
Stanford Univ., Calif.	*	*	16	*	1	*	59
Univ. of California	*	*	8			*	
Univ. of Calif. in Los Angeles	*						
Univ. of Santa Clara, Calif.	*						
Univ. of So. California	*	*	16	*	7	*	6
Colorado State Univ.	*	*	7			*	14
Univ. of Colorado	*	*		*		*	1
Univ. of Denver, Colo.	*						
USAF Academy, Colo.	*						
Yale Univ., Conn.	*	*	18	*	16	*	28
Univ. of Bridgeport, Conn.	*						
Univ. of Connecticut	*	*	12			*	1
Trinity College, Conn.	*						
Univ. of Delaware	*	*	10	*	3	*	32
Catholic Univ. of America, D. C.	*	*		*		*	
George Washington Univ. D. C.	*	*	2	*			
Univ. of Florida	*	*	1	*	2		
Univ. of Miami, Fla.	*	*	2	*	10		
Georgia Inst. of Tech.	*	*		*		*	24
Univ. of Idaho	*					*	3
Bradley Univ., Ill.	*						
Illinois Inst. of Tech.	*	*		*		*	5
Northwestern Univ., Ill.	*	*	5	*	3	*	39
Univ. of Chicago, Ill.	*	*	20	*	10	*	
Univ. of Illinois	*	*	13	*	5	*	37
Notre Dame Univ., Ind.	*	*	43	*	17	*	40
Purdue Univ., Ind.	*	*	5	*	2	*	55
Iowa State College	*	*		*		*	14
Univ. of Iowa	*	*	43	*	43	*	
Kansas State College	*	*	5	*	6		
Univ. of Kansas	*	*	3	*	4	*	6
Univ. of Wichita, Kansas	*			*	4		
Univ. of Kentucky	*	*	1	*		*	
Louisiana Poly. Inst.	*						
Louisiana State Univ.	*	*	26	*	11	*	19
Southwestern Louisiana Inst.	*						
Tulane Univ., La.	*	*	6	*	20	*	11
Univ. of Maine	*					*	1
Johns Hopkins Univ., Md.	*	*	62	*	9	*	74

Educational Institutions (continued)	Partial Programs in Astronautics	Programs in Physics	Fellowships in Physics	Programs in Mathematics	Fellowships in Mathematics	Programs in Engineering	Fellowships in Engineering
Univ. of Maryland	*	*		*	10	*	8
Harvard Univ., Mass.	*	*	18	*	9	*	46
Lowell Tech. Inst., Mass.	*						
Mass. Inst. of Tech.	*	*	48	*	33	*	349
Univ. of Massachusetts	*						
Worcester Poly. Inst., Mass.	*					*	14
Michigan State Univ.	*					*	1
Univ. of Detroit, Mich.	*	*	5	*			
Univ. of Michigan	*	*	8	*	10	*	59
Univ. of Minnesota	*	*		*		*	15
Mississippi State Univ.	*	*					
Univ. of Mississippi	*						
Missouri School of Mines	*						
St. Louis Univ., Mo.	*	*	11	*	13		
Washington Univ. of St. Louis	*	*	4	*		*	9
Montana State College	*					*	2
Univ. of New Hampshire	*					*	1
Princeton Univ., N. J.	*	*	7	*	5	*	37
Rutgers Univ., N. J.	*	*	10	*	4	*	2
Stevens Inst. of Tech., N. J.	*	*		*		*	24
New Mexico Coll. of A. and M. A.	*						
Univ. of New Mexico	*	*	5	*	1		
City Coll. of New York	*						
Clarkson Coll. of Tech., N. Y.	*					*	10
Columbia Univ., N. Y.	*	*		*	1	*	41
Cornell Univ., N. Y.	*	*	3	*	8	*	36
Manhattan College, N. Y	*			*		*	
New York Univ.	*	*		*	2	*	11
Polytechnic Inst. of Brooklyn, N. Y.	*	*	6	*	2	*	22
Rensselaer Poly. Inst., N. Y.	*	*	2	*		*	247
Syracuse Univ., N. Y.	*	*	2			*	2
Univ. of Buffalo, N. Y.	*						
Univ. of Rochester, N. Y.	*	*	35	*	7	*	9
Union College, N. Y.	*						
North Carolina State College	*	*				*	2
Univ. of North Dakota	*					*	4
AF Inst. of Tech., Ohio	*						
Case Inst. of Tech., Ohio	*	*	2	*		*	16
Fenn College, Ohio	*						
Ohio Northern Univ.	*					*	
Ohio Univ.	*					*	1
Ohio State Univ.	*	*	7	*		*	15
Univ. of Akron	*						
Univ. of Cincinnati, Ohio	*	*	8	*	5	*	1
Univ. of Toledo, Ohio	*					*	
Western Reserve Univ., Ohio	*	*					
Oklahoma State Univ.	*			*		*	
Univ. of Oklahoma	*			*			

Educational Institutions (continued)	Partial Programs in Astronautics	Programs in Physics	Fellowships in Physics	Programs in Mathematics	Fellowships in Mathematics	Programs in Engineering	Fellowships in Engineering
Oregon State College	*	*		*		*	6
Carnegie Inst. of Tech., Pa.	*	*	8	*		*	45
Drexel Inst. of Tech., Pa.	*						
Lehigh Univ., Pa.	*	*	1	*	1·	*	25
Penn. Military College	*						
Penn. State Univ.	*	*				*	
Univ. of Pennsylvania	*	*	6			*	10
Univ. of Pittsburgh	*	*				*	
Villanova Univ. Pa.	*						
Brown Univ., R. I.	*	*	4	*	5	*	1
Univ. of Rhode Island	*					*	1
Clemson College, S. C.	*					*	3
South Dakota State Coll.	*						
A. and M. of Texas	*	*	2	*		*	36
Lamar State Coll. of Tech., Texas	*						
Rice Institute , Texas	*	*	19	*	4	*	11
Southern Methodist Univ. , Texas	*			*	2	*	2
Univ. of Houston, Texas	*						
Univ. of Texas	*	*	8	*	4	*	21
Utah State Univ.	*			*	4	*	1
Univ. of Vermont	*						
Virginia Poly. Inst.	*	*				*	23
Seattle Univ., Wash.	*						
State Coll. of Washington	*	*		*			
Univ. of Washington	*	*		*		*	12
West Virginia Univ.	*			*		*	6
Marquette Univ.	*						
Univ. of Wisconsin	*	*	3	*	4	*	10

TABLE D   A Typical 2-year Engineering Curriculum oriented towards Astronautics

Source: *Jets* (these courses were offered by Ohio College of Applied Sciences.

ENGINEERING TECHNOLOGIES

Electrical (power*)

FIRST TERM
  Algebra and Trigonometry I
  English I (written and oral)
  Physics I
  Engineering Drawing I
  Dc Circuits
  Dc Machinery
  General Psychology I

*ALSO AVAILABLE IS ELECTRONICS
OPTION WHICH INCLUDES COURSES IN
TV STUDIO OPERATION, FCC PREPARATION,

Mechanical

FIRST TERM
  Algebra and Trigonometry I
  English I (written and oral)
  Physics I
  Engineering Drawing I
  Mechanical Processes and Equipment
  Electricity (Dc circuits)
  General Psychology I

ELECTRONIC LAB TECHNIQUES, SPECIAL
TUBES AND CIRCUITS, SEMICONDUCTORS,
VIDEO EQUIPMENT AND OPERATION

SECOND TERM
Algebra and Trigonometry II
English II (written and oral)
Physics II
Mechanical Processes and Equipment
Electrical Drafting
Ac Circuits
General Psychology II

THIRD TERM
Introductory Calculus
Machine Processes I

Electronic Drafting
Electronics I
Ac Machinery
Chemistry I

FOURTH TERM
Calculus
Industrial Psychology
Thermodynamics
Strength of Materials
Electronics II
Electrical Design I
Chemistry II

FIFTH TERM
Statistical Mathematics
Industrial Management
Heating and Air Conditioning
Electrical Controls
Electronic Systems
Power Systems
Electrical Design II

SECOND TERM
Algebra and Trigonometry II
English II (written and oral)
Physics II
Manufacturing Processes and
Materials
Engineering Drawing II
Electricity (Dc Machines)
Chemistry I
Physical Metallurgy
General Psychology II

THIRD TERM
Introductory Calculus
Tool Design I
Machine Processes I
Strength of Materials I
Mechanism
Electricity (Ac Circuits)
Chemistry II

FOURTH TERM
Calculus
Industrial Psychology
Machine Processes II
Thermodynamics I
Strength of Materials II
Machine Design I
Electricity (Ac Machines)

FIFTH TERM (Heat Power Option)
Statistical Mathematics
Industrial Management
Heating and Air Conditioning
Thermodynamics II
Internal Combustion Engines
Gas Turbines
Machine Design II (theory)
Heat Power Systems Design

FIFTH TERM (Design Option)
Statistical Mathematics
Industrial Management
Thermodynamics II
Internal Combustion Engines
Gas Turbines
Tool Design II
Machine Design II (theory)
Machine Design II (lab)

The Financial Rewards of Careers in Space

TABLE A

MEDIAN ANNUAL SALARIES OF EMPLOYED SCIENTISTS
by work activity, type of employer, and highest degree, 1960.

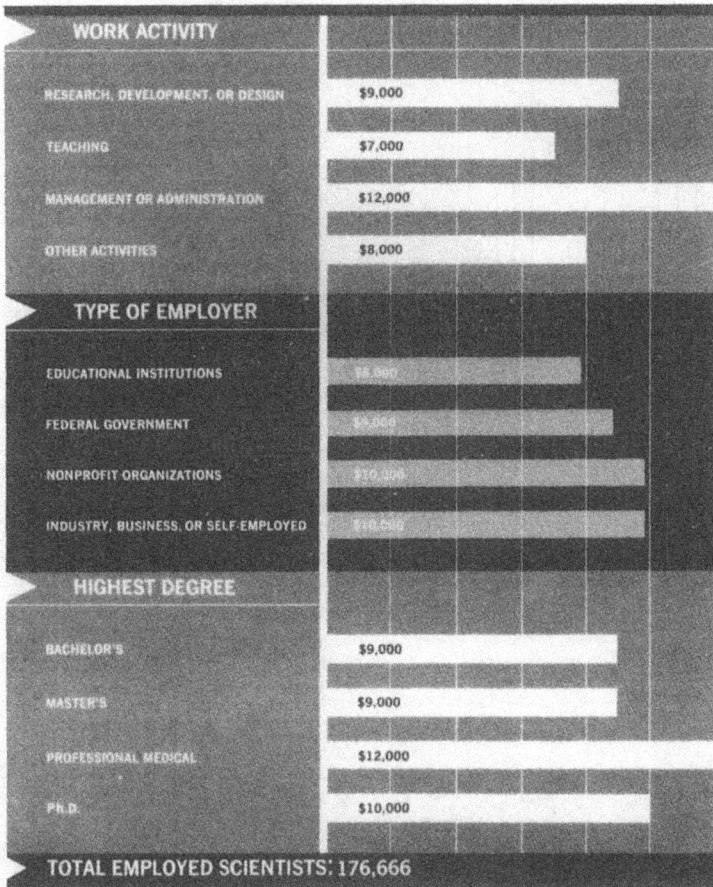

WORK ACTIVITY	
RESEARCH, DEVELOPMENT, OR DESIGN	$9,000
TEACHING	$7,000
MANAGEMENT OR ADMINISTRATION	$12,000
OTHER ACTIVITIES	$8,000
TYPE OF EMPLOYER	
EDUCATIONAL INSTITUTIONS	$8,000
FEDERAL GOVERNMENT	$9,000
NONPROFIT ORGANIZATIONS	$10,000
INDUSTRY, BUSINESS, OR SELF-EMPLOYED	$10,000
HIGHEST DEGREE	
BACHELOR'S	$9,000
MASTER'S	$9,000
PROFESSIONAL MEDICAL	$12,000
Ph.D.	$10,000

TOTAL EMPLOYED SCIENTISTS: 176,666

SOURCE: National Register of Scientific and Technical Personnel, 1960

TABLE B   Source: Engineer Joint Council, *Professional Income of Engineer*, 1960

Salaries of Engineering Graduates Employed in Industry

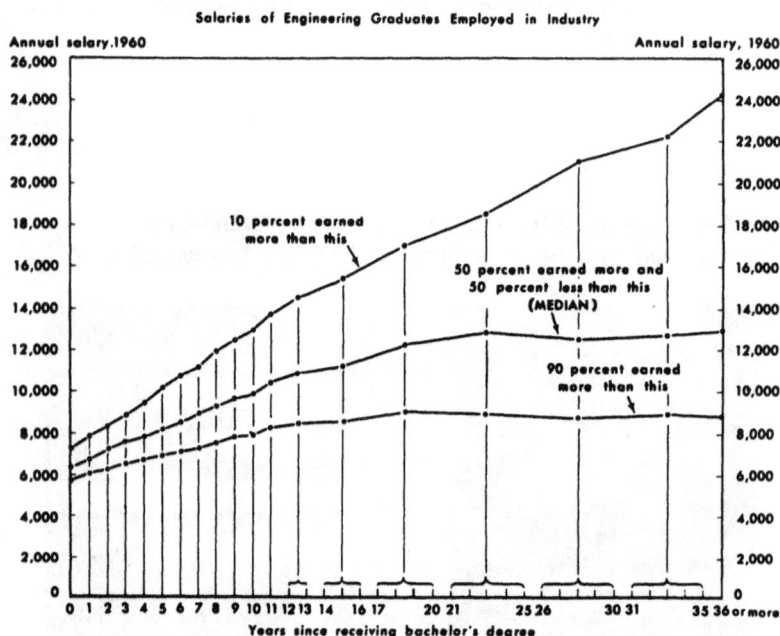

TABLE C   Median Annual Incomes of Men and Women in Mathematical
Employment, by Educational Level, 1960   Source: NSF-62-12

In this computerized age a high premium is placed on mathematical skills,
as this brief table suggests.

Educational level	All respond-ents	Men	Women
All educational levels_	$8, 500	$8, 900	$6, 600
Doctor's degree_ _ _ _ _ _ _ _ _ _ _ _	13, 000	13, 100	11, 000
Master's degree_ _ _ _ _ _ _ _ _ _ _ _	9, 900	10, 100	8, 000
Bachelor's degree_ _ _ _ _ _ _ _ _ _	7, 700	8, 100	6, 500
No degree_ _ _ _ _ _ _ _ _ _ _ _ _ _	7, 900	8, 000	6, 400

TABLE D   Some Typical Starting Salaries in the Field of Electronics

Source: *Radio and TV News*

FIELD	POSITION	INCOME
Electronic Equipment Manufacturers	Design Engineer	7-10,000
	Product Engineer	7-12,000
	Test Engineer: Plan procedures	5- 9,500
	Quality Control Engr.: Supervise inspectors	6- 9,500
	Assistant Engineer	5.5- 8,000
	Junior Engineer	5- 7,000
	Lab Technician: Build and test	4.5- 7,000
	Field Engr.: Coordinate Service	5.5- 8,000
	Air Test Radio Engineer	5- 7,500
	Field Repairman: Maintain and Repair	6- 8,000
	Coil Engineer	6- 9,000
	Transmitter Engineer	6.5-12,000
Radio and TV Station	Transmitter Designer	6-12,000
	Technician: Repair & Maintenance	4.5- 7,000
	Camera Man	7.5-12,000
Airlines	Radio Operator	5.5- 9,000
	Electronic Equipment Engineer	6-10,000
	Radar Installations	6.5-12,000
Business Machine Companies & Computer Firms	Field Service Engr.: Installation	5.5- 9,000 5.5-10,000
Technical Manuals Radio-TV Distributor	Draftsmen	5.5-10,000
	Technical Writer: Prepare data, manuals.	5- 8,000
	Service Technician	5- 7,500
Fire, Police Municipal Governments	Operator: Adjust & Transmit	5.5- 7,500
	Mobile Installation Technician	5- 7,000
Public Utilities	Technician: Microwave maintenance technician	5.5- 7,500
Phone Companies	Equipment maintenance technician	4.5- 8,500
Sales Organization	Sales Engr.: Sells tech. items	6-15,000

TABLE E   Starting Salaries of Engineers and Scientists, 1963

As this book was going to press some fragmentary new data on starting salaries became available. Not only do they show the expected increase over former years, but they also suggest that engineers have begun to outstrip scientists in earning power (at the starting levels, at least). Here are some figures just supplied by The Engineering Manpower Commission of the Engineers Joint Council:

Starting salaries, average (per month)	Scientists	Engineers
for B.S.	$525	$540
for M.S.	$630	$630
for Ph.D.	$845	$870

These figures apply to the general field. In astronautics they are even higher. *Time* magazine reports starting salaries for astronautics engineers with B.S. degrees as high as $650. Engineers with Ph.D.'s command around $1,000.

APPENDIX 3

How and Where Scientists and
Technicians are Employed

TABLE A

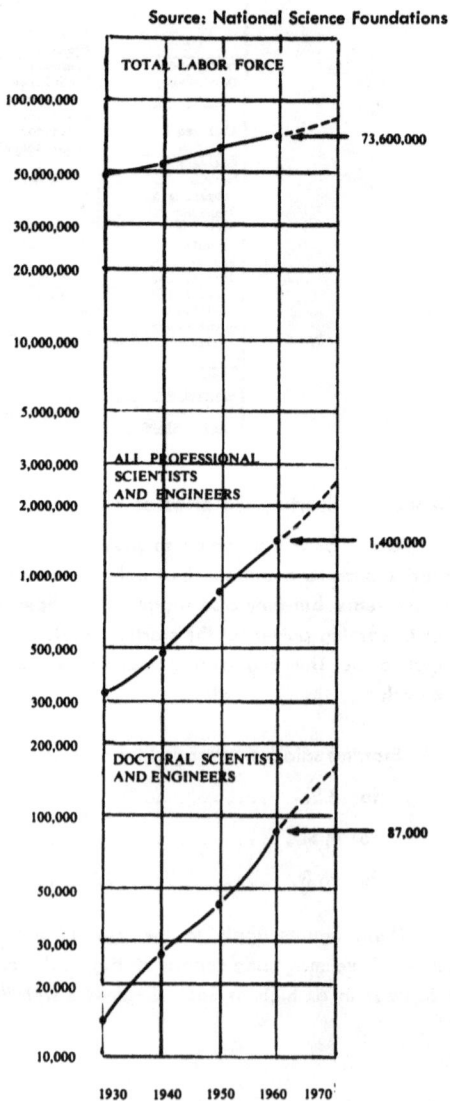

Source: National Science Foundations

TOTAL LABOR FORCE

73,600,000

ALL PROFESSIONAL
SCIENTISTS
AND ENGINEERS

1,400,000

DOCTORAL SCIENTISTS
AND ENGINEERS

87,000

1930   1940   1950   1960   1970

TABLE C

WORK ACTIVITIES OF SCIENTISTS HOLDING BACHELOR'S, MASTER'S, AND Ph.D. DEGREES

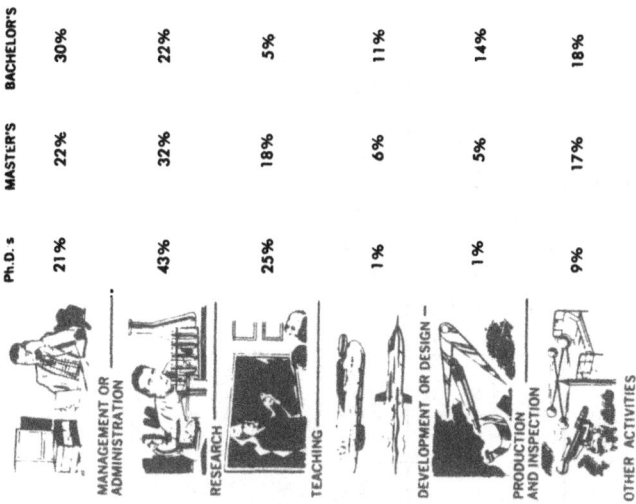

	Ph.D.'s	MASTER'S	BACHELOR'S
MANAGEMENT OR ADMINISTRATION	21%	22%	30%
RESEARCH	43%	32%	22%
TEACHING	25%	18%	5%
DEVELOPMENT OR DESIGN	1%	6%	11%
PRODUCTION AND INSPECTION	1%	5%	14%
OTHER ACTIVITIES	9%	17%	18%

TABLE B

TYPE OF EMPLOYER OF SCIENTISTS HOLDING BACHELOR'S, MASTER'S, AND Ph.D. DEGREES.

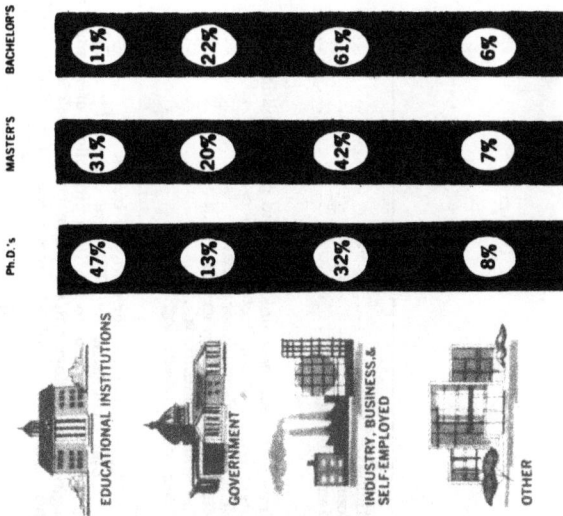

	Ph.D.'s	MASTER'S	BACHELOR'S
EDUCATIONAL INSTITUTIONS	47%	31%	11%
GOVERNMENT	13%	20%	22%
INDUSTRY, BUSINESS, & SELF-EMPLOYED	32%	42%	61%
OTHER	8%	7%	6%

Source: National Register of Scientific and Technical Personnel, 1960

TABLE D   Employment of Technicians in Selected Industries, 1959 and Projected 1970

Source: The National Science Foundation

Industry	1959			Projected 1970 [2]	
	Scientists and engineers	Technicians	Ratio of technicians to scientists and engineers	Scientists and engineers	Technicians [3]
Mining [1]	30,500	8,600	0.28	79,500	22,400
Construction	53,700	21,200	.39	110,600	43,600
Manufacturing	546,000	374,700	.69	1,094,400	760,700
Transportation, communication, and public utilities	59,500	53,300	.90	84,800	76,000
Engineering and architectural services	56,100	62,300	1.11	72,100	80,500
Medical and dental laboratories	1,000	6,400	6.47	1,300	8,500
Miscellaneous business services	7,000	4,400	.63	12,300	7,800
Nonprofit organizations	[4] 6,000	....	....	7,900	....
Other nonmanufacturing	26,300	18,500	.70	52,700	37,000

[1] Without estimates for company births.

[2] Total technicians for 1970 were derived by applying the 1959 technician ratios to the 1970 projections of scientific and engineering employment, except for the chemicals industry. In this industry, the 1970 ratio was adjusted from 0.41 to 0.46 on the basis of information obtained in the special industry studies.

[3] Projections of scientific, engineering, and technician employment for all manufacturing and for the chemicals and electrical equipment industries includes adjustments made on the basis of information obtained in the special industry studies. Unadjusted 1970 projections are as follows:

	Scientists and engineers	Ratio of technicians to scientists and engineers	Technicians
Manufacturing	1,133,400	0.69	782,100
Chemicals	151,500	.41	62,700
Electrical equipment	250,000	.73	181,900

[4] Data are for January 1958.

NOTE.—Totals and ratios are calculated on the basis of all significant digits and therefore may not correspond exactly with those indicated by the rounded figures shown.

TABLE E   The Number of R & D Scientists and Engineers Employed
Full-time, by Industry, January 1962.

Source: The National Science Foundation

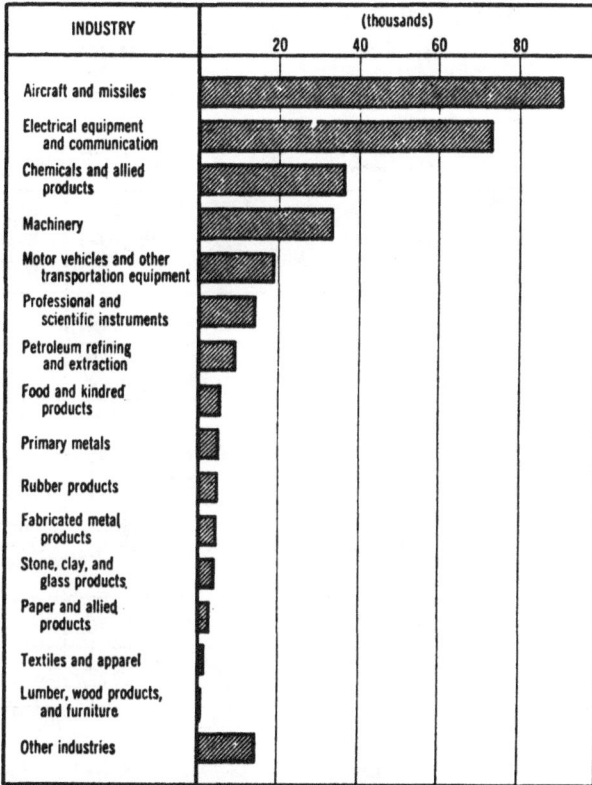

INDUSTRY	(thousands) 20 · 40 · 60 · 80
Aircraft and missiles	
Electrical equipment and communication	
Chemicals and allied products	
Machinery	
Motor vehicles and other transportation equipment	
Professional and scientific instruments	
Petroleum refining and extraction	
Food and kindred products	
Primary metals	
Rubber products	
Fabricated metal products	
Stone, clay, and glass products	
Paper and allied products	
Textiles and apparel	
Lumber, wood products, and furniture	
Other industries	

TABLE F   Employment of Scientists and Engineers in Government and
Educational Institutions.   Source: The National Science Foundation

*—Employment of scientists and engineers in colleges and
universities, by occupation, 1959 and projected 1970*

Occupation	1959	Projected 1970	Percent change, 1959–70
Total............	112, 300	176, 400	57. 2
Engineers..............	24, 200	38, 100	57. 2
Scientists..............	88, 000	138, 300	57. 1

*—Employment of scientists and engineers in the Federal
Government, by occupation, 1959 and projected 1970*

Occupation	1959	Projected 1970	Percent change, 1959–70
Total............	101, 400	164, 300	62. 0
Engineers..............	53, 900	77, 700	44. 3
Scientists..............	47, 500	86, 600	82. 2

*—Employment of scientists and engineers in State
governments, by occupation, 1959 and projected 1970*

Occupation	1959	Projected 1970	Percent change, 1959–70
Total...........	39, 400	73, 100	85. 3
Engineers..............	28, 200	52, 200	85. 3
Scientists..............	11, 300	20, 900	85. 4

*—Employment of scientists and engineers in local govern-
ments, by occupation, 1959 and projected 1970*

Occupation	1959	Projected 1970	Percent change, 1959–70
Total............	30, 000	49, 600	65. 3
Engineers..............	24, 000	39, 700	65. 3
Scientists..............	6, 000	9, 900	65. 3

TABLE G   Aero-Space Specialties at the Principal NASA Research
Centers   Source: NASA

For *Ames Research Center*, send your application and any attachments to:

> Board of U.S. Civil Service Examiners
> NASA Ames Research Center
> Mountain View, Calif.

For *Flight Research Center* or *Western Operations Office*, send your application and any attachments to:

> Board of U.S. Civil Service Examiners
> NASA Flight Research Center
> P.O. Box 273, Edwards, Calif.

For *Goddard Space Flight Center* or *NASA
Headquarters*, send your application and any attachments to:

> Board of U.S. Civil Service Examiners
> NASA Goddard Space Flight Center
> Greenbelt, Md.

For *Langley Research Center* or *Wallops Space
Flight Station*, send your application and any
attachments to:

> Board of U.S. Civil Service Examiners
> NASA Langley Research Center
> Hampton, Va.

For *Lewis Research Center*, send your application
and any attachments to:

> Board of U.S. Civil Service Examiners
> NASA Lewis Research Center
> Cleveland 35, Ohio

For *Marshall Space Flight Center* or *Launch
Operations Directorate, Cape Canaveral, Fla.* or
*Army Ordnance Missile Command, Redstone
Arsenal, Ala.*, send your application and any at-
tachments to:

> Board of U.S. Civil Service Examiners
> NASA Marshall Space Flight Center
> Huntsville, Ala.

For *positions from grade GS–16, $15,255 up to $21,000 a year* (see page 14), interested and qualified persons should make separate inquiry to:

Director of Personnel
National Aeronautics and Space Administration
Washington 25, D.C.

Listed on the following page are the Aero-Space Technology specialties in physical sciences, engineering, and mathematics. Each listed specialty is described in a preceding section on the pages referenced below. The X-marks in the table indicate which specialties are of major importance at the NASA Centers. Each Center may have some work or a few individuals in additional specialties.

This table, including the information immediately following, presents the types and locations of NASA work in physical sciences, engineering, and mathematics as of the date of this announcement. But it is subject to changes in the specialties due to the rapidly changing frontiers of aero-space technology, and changes in locations of the work due to shifts in functions or the development of new experimental facilities, without published amendment.

See table on next page.

	Ames Research Center	Flight Research Center	Goddard Space Flight Center	Langley Research Center	Lewis Research Center	Marshall Space Flight Center
Fluid and Flight Mechanics						
Flight Mechanics	X	X	X	X	X	X
Fluid Dynamics	X	X	X	X	X	X
Magnetofluiddynamics	X	X		X	X	X
Aerostructural Dynamics	X			X		X
Flight Vehicle Acoustics		X		X	X	X
Propulsion System Dynamics		X		X	X	X
Heat Transfer	X	X	X	X	X	X
Stability and Control	X	X		X	X	X
Control and Guidance Systems	X	X	X	X	X	X
Trajectory-Orbit Analysis	X	X	X	X		X
Energy and Power Systems						
Nuclear Propulsion and Power			X		X	X
Electrical Propulsion and Power					X	X
Chemical Propulsion and Power				X	X	X
Internal Flow Dynamics				X	X	X
Friction and Lubrication					X	X
Materials and Structures						
Structural Mechanics	X	X		X	X	X
Flight Structures		X		X	X	X
Structural Materials			X	X		X
Basic Properties of Materials				X	X	X

	Ames Research Center	Flight Research Center	Goddard Space Flight Center	Langley Research Center	Lewis Research Center	Marshall Space Flight Center
Space Sciences	X	X	X			
Aeronomy	X	X	X			
Ionospheric Studies			X			X
Particles and Fields			X			X
Stellar and Galactic Studies			X			
Lunar and Planetary Studies	X		X			
Measurement and Control Systems	X	X	X	X	X	X
Sensors and Transducers	X	X	X	X	X	X
Heat and Light Measurement	X	X	X	X	X	X
Radio Frequency Systems	X	X	X	X		X
Force Measurement	X	X		X		X
Measurement Standards and Calibration	X	X	X	X		X
Automated Control Systems	X	X	X	X	X	X
Data Systems	X	X	X	X	X	X
Theoretical Computation Techniques	X	X	X	X	X	X
Theoretical Simulation Techniques	X	X	X	X	X	X
Experimental Facilities and Techniques	X	X	X	X	X	X
Experimental Facilities	X		X	X	X	X

Experimental Equipment	Fluid and Flow Systems	Electrical Experimental Equipment	Experimental Facility Techniques	Nuclear Facility Techniques	Flight and Launch Vehicle Systems	Launching Systems and Techniques	Flight Vehicle Experimental Techniques
X	X	X	X	X	X	X	X
X	X	X	X	X			
X	X	X	X		X	X	
X			X				
					X	X	X
X	X	X					

APPENDIX 4
Sales Data on the Major Aerospace Firms

PART I. THE LEADING FIRMS

Unlike the case with other major industries, such as automobiles or steel products, the precise total sales of aerospace products (jetcraft and rocketry) are not always available from firms or government sources. Among the aircraft-makers who have partly turned to missiles, the breakdown between jet and rocket production is not generally obtainable. Hence the following statistics are not complete, and estimates are substituted where exact figures are nonexistent. Despite these handicaps, the writer feels it important to present (possibly for the first time in public channels) as many data as are available, in order to give science students a fairly reliable picture of the aerospace industry as a whole.

Certain definitions are important. *Gross* sales for each firm include *all* their products, space or non-space. The *aerospace* portion of that gross means jetcraft plus missiles plus space hardware. In turn, the *missile/space* portion of the aerospace total (seldom defined except in broad percentages) excludes jets and other aircraft, and narrows down to DOD's military missiles and NASA's space vehicles. Finally, there is a rough breakdown between military missiles and space rocketry, by virtue of DOD and NASA funding totals.

Note that in the list that follows, *FY-1962 Aerospace* figures refer to contracts placed for future fulfillment, and have no direct bearing on the firm's CY-1962 gross sales. Aerospace funding of 1962 will show up as income mainly in a firm's 1963 records. Also note that most of the air firms (either airplane-makers or heavily committed suppliers) suffered a drop in sales about 1960. This was due, not to poor business methods, but to cutbacks DOD began making in jetcraft at that time, as the nation began switching decisively from the Air Age to the Missile Age. And this was before military missile contracts could pile up and fill the gap left by "sudden-death" aircraft contracts. However, by 1961-62, most of these aircraft-oriented companies had picked up enough new military rocket work from DOD, and from NASA's rapidly growing funds for space hardware, to again show rising curves for gross sales. These curves are likely to rise spectacularly from 1963 on, as military rocketry production and the civilian drive into space gain momentum. Most firms will soon share the spectacular gross-sales jumps already displayed by a few key firms—North American, Aerojet-General and Northrop, for example.

Finally, note that the first figure in parentheses, preceding each firm, indicates its position among the Top Hundred Aerospace Aces, according to "anticipated" new income in 1963 coming out of combined DOD and NASA contracting—or strictly missiles/space business—as of fiscal 1962. These government funds for rockets and space vehicles—which can only *increase* as time goes on—therefore represent an "astronautics barometer" as to which firms will benefit most from "space prosperity" through this decade to 1970—and beyond.

Chapter 9

(2) North American Aviation Inc. *Gross:* CY-1960, $964,162,000; CY-1961, $1,262,333,000; CY-1962, $1,633,675,000; CY-1963 (est.), $1,950,000,000. *FY-1962 Aerospace:* DOD, $1,032,500,000; NASA, $199,109,000; Total, $1,231,609,000.

(11) Douglas Aircraft Co. Inc. *Gross:* CY-1960, $1,174,041,000; CY-1961, $791,312,000; CY-1962, $1,200,000,000; CY-1963 (est.), $1,350,000,000 (missiles/space = 35%). *FY-1962 Aerospace:* DOD, $365,600,000; NASA, $68,374,000; Total, $433,974,000.

(4) The Boeing Co. *Gross:* CY-1960, $1,554,600,000; CY-1961, $1,800,900,000; CY-1962, $1,768,500,000; CY-1963, $2,250,000,000. *FY-1962 Aerospace:* DOD, $1,132,800,000; NASA, $15,584,000; Total, $1,148,384,000.

(24) Chrysler Corp. *Gross:* CY-1962, $2,378,000,000 (missiles/space = 7%, but rising rapidly). *FY-1962 Aerospace:* DOD, $181,500,000; NASA, $31,294,000; Total, $212,794,000. (Note: Potential gross of Saturn booster contracts *only*, from 1962 to 1970, will be in the billion-dollar-plus range.)

Chapter 10

(6) Martin-Marietta Corp. *Gross:* CY-1960, $1,019,300,000; CY-1961, $1,213,100,000; CY-1962, $1,199,574,000; CY-1963 (est.), $1,250,000,000 (missiles/space = 67%). *FY-1962 Aerospace:* DOD $802,700,000; NASA, $1,844,000; Total, $804,544,000.

(5) General Electric Co. *Gross:* CY-1962, $4,792,700,000. *Sub-Gross* (jets/missiles/space only): CY-1960, $1,049,400,000; CY-1961, $1,213,100,000; CY-1962, $1,377,900,000; CY-1963 (est.), $1,500,000,000. *FY-1962 Aerospace:* DOD, $975,900,000; NASA, $22,995,000; Total, $998,895,000.

(1) Lockheed Aircraft Corp. *Gross:* CY-1960, $1,332,200,000; CY-1961, $1,444,510,000; CY-1962, $1,753,074,000; CY-1963 (est.), $1,900,000,000 (missiles/space = 50%). *FY-1962 Aerospace:* DOD, $1,419,500,000; NASA, $4,951,000; Total, $1,424,451,000.

(3) General Dynamics Corp. *Gross:* CY-1960, $1,987,749,000; CY-1961, $2,062,378,000; CY-1962, $1,898,500,000; CY-1963 (est.), $2,225,000,000. *FY-1962 Aerospace:* DOD, $1,196,100,000; NASA, $27,937,000; Total, $1,224,037,000.

(12) Aerojet-General Corp. *Gross:* CY-1960, $425,101,000; CY-1961, $479,402,000; CY-1962, $604,857,000; CY-1963 (est.), $750,000,000. (Note: Figures include other aerospace divisions of the parent firm, General Tire & Rubber Co. Gross of GT&R for FY-1962: $959,770,000.) *FY-1962 Aerospace:* DOD, $364,100,000; NASA, $67,314,000; Total, $431,414,000.

(7) United Aircraft Corp. *Gross:* CY-1960, $987,880,000; CY-1961, $1,094,756,000; CY-1962, $1,160,459,000; CY-1963 (est.), $1,250,000,000 (missiles/space = 10%). *FY-1962 Aerospace:* DOD, $662,700,000; NASA, $34,057,000; Total, $696,757,000.

(14) McDonnell Aircraft Corp. *Gross:* CY-1960, $436,981,000; CY-1961, $344,414,000; CY-1962, $390,718,000; CY-1963 (est.), $525,000,000 (missiles/space = 25%). *FY-1962 Aerospace:* DOD, $310,900,000; NASA, $68,470,000; Total, $379,370,000.

(29) Thiokol Chemical Corp. *Gross:* CY-1960, $171,540,000; CY-1961, $193,900,000; CY-1962, $255,808,000; CY-1963 (est.), $300,000,000. *FY-1962 Aerospace:* DOD, $178,300,000; NASA, $849,000; Total, $179,149,000.

(31) Thompson-Ramo-Wooldridge Inc. *Gross:* CY-1960, $420,-400,000; CY-1961, $409,000,000; CY-1962, $460,314,000; CY-1963 (est.), $475,000,000 (missiles/space = 40%). *FY-1962 Aerospace:* DOD, $148,600,000; NASA, $17,085,000; Total, $165,685,000.

(15) Radio Corporation of America. *Gross:* CY-1961, $1,545,-912,000; CY-1962, $1,751,646,000; CY-1963 (est.), $1,950,000,000. *FY-1962 Aerospace:* DOD, $339,600,000; NASA, $20,187,000; Total, $359,787,000.

Chapter 11

(23) Hughes Aircraft Co. *Gross:* Company does not release these figures to the public. *FY-1962 Aerospace:* DOD, $234,200,000; NASA, $9,178,000; Total, $243,378,000.

(17) Grumman Aircraft Engineering Corp. *Gross:* CY-1960, $325,500,000; CY-1961, $316,700,000; CY-1962 (est.), $325,000,-000; CY-1963 (est.), $375,000,000 (missiles/space = 15% today, but percentage is rising rapidly with OAO and LEM contracts from NASA). *FY-1962 Aerospace:* DOD, $303,600,000; NASA, $24,586,-000; Total, $328,186,000.

(?) B. F. Goodrich Co. (Because this firm is one of the few giants engaged mainly in *sub-contracting* for the primes, it shows little direct DOD funding, but will continue to get at least 30% of its gross from aerospace products.) *Gross:* CY-1960, $764,736,000; CY-1961, $757,785,000; CY-1962, $812,026,000; CY-1963 (est.), $900,-000,000 (missiles/space = 25%). *FY-1962 Aerospace:* DOD, $417,-000; NASA, $593,000; Non-direct DOD/NASA (est.), $250,000,000. (Unofficial rating among Top Hundred = 25th.)

(20) Ford Motor Co. *Gross:* CY-1962, $8,089,617,000. *Sub-Gross* (aerospace only, excluding autos, as per Aeronutronic and Philco Divisions): CY-1960, $125,700,000; CY-1961, $200,800,000; CY-1962, $273,479,000; CY-1963 (est.), $350,000,000. *FY-1962 Aerospace:* DOD, $269,100,000; NASA, $4,379,000; Total, $273,479,000.

(101) Atlantic Research Corp. (Representing the lower-funded but still vitally important group of sub-prime/sub-contracting firms below the Top Hundred. In this case, its true position is not reflected by its government-funding sums, but rather by the firm's *gross*—which is *entirely* in Space Age products and services. And note the firm's spectacular growth curve each year.) *Gross:* CY-1960, $13,513,-000; CY-1961, $20,362,800; CY-1962 (est.), $36,000,000; CY-1963 (est.), $50,000,000. *FY-1962 Aerospace:* $8,525,000 (DOD only, advanced research).

(18) Avco Corp. *Gross:* CY-1960, $322,750,000; CY-1961, $323,150,000; CY-1962, $414,280,000; CY-1963 (est.), $500,000,-000 (missiles/space = 35%). *FY-1962 Aerospace:* DOD, $323,300,-000; NASA, $1,447,000; Total, $323,747,000.

(74) Marquardt Corp. *Gross:* CY-1960, $66,000,000; CY-1961, $48,600,000; CY-1962, $50,600,000; CY-1963 (est.), $60,000,000 (missiles/space = almost 100%). *FY-1962 Aerospace:* DOD (all), $38,800,000. (Note: This firm's all-space gross brings its true rating up to about 50th.)

(34) Northrop Corp. *Gross:* CY-1960, $234,000,000; CY-1961, $268,000,000; CY-1962, $348,300,000; CY-1963 (est.), $375,000,-000 (missiles/space = 60%). *FY-1962 Aerospace:* DOD, $152,500,-000; NASA, $1,329,900; Total, $154,536,000.

(8) Bell Telephone Laboratories Inc. (plus Western Electric Co.; divisions of AT&T). *Gross:* CY-1961, $634,000,000; CY-1962, $542,000,000. *FY-1962 Aerospace* (BTL & Western Electric): DOD, $467,700,000; NASA, $10,833,000; Total, $478,533,000.

(13) Raytheon Co. *Gross:* CY-1960, $539,900,000; CY-1961, $562,900,000; CY-1962, $580,721,000; CY-1963 (est.), $600,000,-000. *FY-1962 Aerospace:* DOD (all), $405,000,000.

(65) Garrett Corp. *Gross:* CY-1960, $223,824,000; CY-1961, $190,975,000; CY-1962, $215,050,000; CY-1963 (est.), $250,000,-000 (missiles/space = 20%). *FY-1962 Aerospace:* DOD (all), $46,-700,000. (NASA funding, which dropped off with the finish of the Mercury Program, will resume in much greater amounts in fiscal 1963 as the Gemini Program—for which Garrett-AiResearch supplies the life-support systems—goes into full swing.)

(37) Minneapolis-Honeywell Regulator Co. *Gross:* CY-1960, $426,100,000; CY-1961, $470,182,000; CY-1962, $596,267,000; CY-1963 (est.), $750,000,000 (missiles/space = 40%). *FY-1962 Aerospace:* DOD, $127,500,000; NASA, $4,681,000; Total, $132,181,000.

PART II—OTHER IMPORTANT FIRMS

These firms complete the list of the Top Aerospace Hundred, according to total FY-1962 contracts awarded by NASA and DOD out of this missile/space budget. Firms reviewed previously are not included. The coded letters and numbers in the firm entries mean the following:

(00) Number indicating firm's position in FY-1962, per total dollar-volume from NASA and DOD combined.

DIV List of divisions, departments or subsidiaries engaged mainly in aerospace work.

T/P Total personnel of corporation.

T/GR Parent corporation's total gross calendar-year sales, including non-aerospace products. (Unless otherwise stated, estimates are the author's, based on the firm's growth-pattern in previous years.)

T/MS Gross of missiles/aerospace only for fiscal 1962 (July 1, 1961, to June 30, 1962). Missiles/space gross (excluding jetcraft) varies between 50% and 75% on the average. (Note: FY-1962 contract income—often higher than CY-1962 gross—will mainly enter CY-1963 gross of firm.)

M/S Firm's main products and/or services in astronautics field (excluding all air-breathing jetplanes and aircraft).

C/P Major contracts and projects currently held in missiles/space field.

(Note: Any item omitted indicates figures were not available.)

(9) *Sperry/Rand Corp.*, 30 Rockefeller Plaza, New York 20, N. Y. DIV: Ford Instrument, Univac, Sperry Electronic Tube, Sperry Gyroscope, Sperry Farragut, Sperry Microwave, Sperry Utah, Vickers Inc. TGR FY-1962: $1,182,555,000 (19% aerospace).

T/MS: $467,760,000 (NASA, $2,160,000; DOD, $465,600,000). M/S: Electronic components in guidance, telemetry, micro-circuits, computers. C/P: Prime for Sergeant Missile. Sub-prime for guidance or other electronic devices in Apollo craft, X-15 rocketplane, Polaris, Titan, Pershing, other military missiles.

(10) *General Motors Corp.*, GM Building, Detroit 2, Mich. DIV: AC Spark Plug, Allison, Defense Systems, Defense Research Labs (5500 scientists and engineers), Delco Radio, New Departure, Packard Electric. T/P: 600,000. T/GR (non-aerospace only, CY-1962): $14,159,322,000. T/GR (missiles and aerospace only): CY-1962, $480,919,000; CY-1963 (est.), $525,000,000. T/MS (FY-1962): $450,417,000 (NASA, $1,417,000; DOD, $449,000,000). M/S: Hyper-ballistics, rocket propulsion accessories, widespread electronics, GSE for missiles, solar and nuclear power-conversion, motorized systems in rocketry transportation and launching. C/P: Inertial guidance sub-prime for Titan, Thor, Mace, Matador, MMRBM; for Apollo spacecraft and varied satellites. Heavy rolling stock for military missiles. Prime for first-stage steel engine casings of Minuteman ICBM, and second-stage titanium casings.

(16) *Republic Aviation Corp.*, Conklin St., Farmingdale, L. I., N. Y. DIV: None. T/GR: CY-1961, $285,000,000; CY-1962, $295,-766,000; CY-1963 (est.), $310,000,000. T/MS (FY-1962): $339,-191,000 (NASA, $6,391,000; DOD, $332,800,000). M/S: Spacecraft battery and power systems, radiation sensors, varied electronics, life-support components for manned craft. C/P: Onboard power-units for many military missiles and space vehicles; variety of electronic gadgets; advanced space studies in lunar/interplanetary manned camps for NASA.

(19) *The Bendix Corp.*, Fischer Building, Detroit 2, Mich. DIV: Bendix-Computer, Bendix-Aerospace, Bendix-Systems, Bendix-Pacific, Eclipse-Pioneer, Friez Instrument, Red Bank. T/GR (firm's 1962 FY): $794,000,000 (missiles/space = 29%). T/MS (FY-1962): $305,312,000 (NASA, $19,412,000; DOD, $285,900,000). M/S: Computers, guidance, servo-mechanisms, satellite instrumentation, launch-pad GSE, electronic circuit systems, tracking radars. C/P: Prime for USN's Typhon, Talos, sub-contracts for Army's Pershing, Hawk, many other missiles.

(21) *Westinghouse Electric Corp.*, 3 Gateway Center, P.O. Box 2278, Pittsburgh 30, Pa. DIV: Aerospace Electrical, Air Arms, Astroelectronics, ¡Astronuclear, Electronic Tube, Micarta Plastics, Semi-conductors, Space Materials, Sunnyvale Mfg. T/P: 125,000. T/GR: CY-1961, $1,913,770,000; CY-1962, $1,954,480,000; CY-

1963 (est.), $2,000,000,000 (missiles/space = 21%). T/MS (FY-1962): $249,433,000 (NASA, $3,433,000; DOD, $246,000,000). M/S: Space power systems, guidance, computers, nuclear propulsion, solid-state semi-conductors, electrical GSE, Space Age metals and plastics, solid rocket casings, military missile mobile launchers. C/P: Prime for USN's Astor, sub-prime for Gemini and Apollo guidance and/or onboard power systems. Guidance devices in Rover nuclear rocket, OAO, X-20, SNAP Project, NERVA Program, S-52 satellite, Polaris, Typhon, Titan-3, Bomarc, Minuteman.

(22) *International Telephone & Telegraph Corp.*, 320 Park Ave., New York 22, N. Y. DIV: ITT Federal Laboratories, ITT Industrial Products, ITT Kellogg, Jennings Radio Co., Surprenant Mfg. Co. T/P: 165,000. T/GR: CY-1962, $1,090,200,000; CY-1963 (est.), $1,250,000,000. T/MS (FY-1962) $245,807,000 (NASA, $2,207,-000; DOD, $243,600,000). M/S: Varied aerospace electronic devices, ComSat design, space simulation systems, telemetry tape, compact computers, space signaling methods. C/P: Sub-prime for SD-2 RATO-boosted missile, also Lacrosse; tracking and reception system for Relay Communications Satellite; widespread sub-contracting in new military rockets (Triton, Meteor) and a dozen operational types; broadband electronic components for aerospace vehicles.

(25) *American Machine & Foundry Co.*, 261 Madison Ave., New York 16, N. Y. DIV: General Engineering Mechanics Research, AMF Atomics. T/MS (FY-1962): $187,300,000 (all DOD). M/S: GSE for launching pads and military missiles, space environment systems, astronaut training equipment, nuclear rocket research equipment, field apparatus and maintenance GSE for rockets. C/P: GSE at Cape Canaveral and PMR, astronaut testing systems at USAF/NASA training centers, safety handling techniques for Project Rover nuclear rocket engines, silo-launch GSE at Atlas and Titan ICBM missile sites.

(26) *Newport News Shipbuilding & Drydock Co.*, Newport News, Va. DIV: Various inland facilities and warehouses. T/MS (FY-1962): $185,000,000 (all DOD). M/S: Water transport systems for large aerospace vehicles and crated components, canal and riverway shippage, offshore launching range equipment. C/P: Ocean and river shipments of aerospace products from California plants to desired assembly and test sites or launch locations; prime for shallow-water astronaut rescue boats and amphibian pick-up craft.

(27) *Hercules Powder Co.*, 910 Market St., Wilmington 99, Del. DIV: Alleghany Ballistic Laboratory. T/P: 30,000. T/GR: CY-1961, $380,200,000; CY-1962, $454,829,000; CY-1963 (est.), $500,-

000,000. T/MS (FY-1962): $181,600,000 (all DOD). M/S: Solid-fuel rocket motors, plastic casings, fuzing, upper stages. Guide propulsion units. C/P: Ranger moon probes, USAF's Blue Scout Junior solid-fuel launcher, Antares and Altair stages of NASA's Scout rocket. Propulsion units in Minuteman, Polaris, Honest John, Little John, Nike Hercules, Talos, Thor-Ablestar, Thor-Delta, Terrier.

(28) *Standard Oil Company* (New Jersey), 30 Rockefeller Plaza, New York, N. Y. T/MS (FY-1962): $180,100,000 (all DOD). M/S: Mainly liquid rocket fuels. Esso supplies RP-1 (kerosene), alcohol, hydrazine and such, which are purchased through DOD since the USAF maintains all present-day launch boosters—Atlas, Thor, Titan, H-1 Saturn engines—even though many space shots are in NASA's program.

(30) *International Business Machines Corp.*, 590 Madison Ave., New York, N. Y. DIV: Federal Systems. T/P: 132,000. T/GR: CY-1961, $1,694,300,000; CY-1962, $1,925,200,000; CY-1963 (est.), $2,200,000,000 T/MS (FY-1962); $168,134,000 (NASA, $12,634,-000; DOD, $155,500,000). M/S: Computers and data-processors for flying hardware, programming sequences for ballistic or space trajectories, guidance GSE data-transmissions. C/P: Sub-prime for Bomarc A&B, OAO, and Gemini. Many launch-pad complexes, ground-command vehicle control systems, spacecraft, ballistic missiles and military rockets (for NASA and USAF/USN/Army) with IBM equipment from the simplest to the most complex types. The IBM 7090 is widely used by aerospace firms for computer and data-processing work, including simulated Saturn flights at MFSC in preflight preliminaries.

(32) *Ling-Temco-Vought Inc.*, Box 5003, Dallas 22, Texas. DIV: Aerospace Systems, Aeronautics & Missiles, Astronautics, Range Systems, Temco Aerosystems, Temco Electronics, Display Systems, Continental Electronics, Ling Electronics, LTV Research Center. T/P: 22,500. T/GR: CY-1961, $193,700,000; CY-1962, $325,439,000; CY-1963 (est.), $400,000,000. T/MS (FY-1962): $160,442,000 (NASA, $27,042,000; DOD, $133,400,000). M/S: Electronics, space simulation tests, missile check-out equipment. C/P: Electronic components and check-out of Scout, Blue Scout, Regulus I and II, and Titan. Space-environment tests of Scout, Titan-2, Saturn, Zuni, Minuteman and X-20.

(33) *FMC Corporation*, 161 E. 42d St., New York 17, N. Y. DIV: None. T/MS (FY-1962): $160,400,000 (all DOD). M/S: Suppliers of hydrogen peroxide for space vehicle jet-controls, and of various specialized rocket propellants (hydrazine, nitrogen tetroxide,

etc.); also heavy-duty missile-support GSE. C/P: Hypergolic (self-igniting) propellants of Titan-2; attitude jet-control fuel for various spacecraft; Bomarc erector-launcher; Mauler missile's auxiliary power unit; tracked transport field vehicles for Pershing.

(35) *Collins Radio Co.*, 5225 C Ave., N.E., Cedar Rapids, Iowa. DIV: Communications Data, Components, Information Science. T/MS (FY-1962): T/GR: CY-1962, $207,775,600; CY-1963 (est.), $275,000,000 (aerospace: 25%). T/MS (FY-1962): $153,769,000 (NASA, $3,669,000; DOD, $150,100,000). M/S: Electronics telemetry, guidance, space communications. C/P: Gemini and Apollo telemetry/FM-HF voice-radio sub-contracting; guidance and command-radio in variety of military missile systems.

(36) *Pan American World Airways Inc.* DIV: Guided Missiles Range, P.O. Box 4187, AFMTC, Patrick AFB, Fla. T/MS (FY-1962): $146,700,000 (all DOD/USAF). M/S: General management of Atlantic Missile Range for the USAF (does little with rocket hardware). C/P: Services involved in maintaining AMR at Cape Canaveral, integrating all tracking chains through the West Indies during satellite launches and manned spaceflights, and complete logistics for launch countdowns.

(37) *Curtiss-Wright Corporation*, 304 Valley Blvd., Woodridge, N. J. DIV: Curtiss, Electronics, Abrams Instrument, Electronic Fittings, Marquette, Wright Aeronautical. T/P: 17,500. T/GR: CY-1961, $203,488,000; CY-1962, $228,726,000; CY-1963 (est.), $245,-000,000. T/MS (FY-1962): $144,600,000 (all DOD). M/S: Radars, computers, stereo space cameras, precision satellite-shell castings, auxiliary rocket propulsion systems. C/P: Sub-prime for Big Solids rocket-engine casings for Titan-3 boosters; many assorted devices in satellites, probes, manned vehicles, military missiles. Quality-control units in manufacturing of Apollo craft and Saturn boosters. Unique new machine-tools for Space Age purposes (notably lightweight precision gears for rocket-making machinery).

(39) *Textron Inc.*, Providence, R. I. DIV: Bell Aerospace Corp., Bell Aerosystems Co., Buffalo 5, N. Y. (these two dominate parent corporation's business at present), Aerospace Rockets, Avionics, Hydraulic Research. T/P: 33,000 (1750 technical: 1000 engineers and scientists, 750 technicians). T/GR (parent corporation in full): CY-1962, $550,000,000; CY-1963, $600,000,000 (est.). T/GR (Bell group only): CY-1961, $54,999,000; CY-1962, $54,931,000; CY-1963 (est. by firm), $80,000,000. T/MS (FY-1962), $118,645,000 (NASA, $1,245,000; DOD, $117,400,000). M/S: Liquid-fuel rocket engines, attitude jet-control systems, guidance servo-mechanisms,

space-pilot flight simulators, advanced astro-research. C/P: Prime for rocket engines of all Agena upper-stages (placing over 50 satellites in orbit to date, including Discoverer series); sub-prime in moon program for LEM lunar-takeoff propulsion unit, and for attitude system of Apollo spacecraft series. Prime for "rock-and-roll" astronaut test-device; also future zero-g "Rocket Belt" for spacemen leaving craft to maneuver in free space. Accelerometers in Midas, Ranger, Skybolt and Sergeant.

(40) *General Telephone & Electronics Corporation*, 730 Third Ave., New York 17, N. Y. DIV: General Tel & Electronics Lab, Automatic Electric, Lenkurt Electric, Sylvania Electric Products (the last handles bulk of aerospace work). T/MS (FY-1962): $116,-300,000 (all DOD). M/S: Complete military electronic systems for missiles; wide line of semi-conductors and microwave components. C/P: R&D of FABMDS for Army (Field Anti-Ballistic Mobile Defense System); sub-prime for MMRBM command-radio; microwave servo-electronics for other missiles.

(41) *Standard Oil Co.* (of Calif.), 225 Bush St., San Francisco 20, Calif. T/MS (FY-1962): $115,200,000 (all DOD). M/S and C/P: Mainly liquid rocket fuels.

(42) *Texaco Inc.*, 135 E. 42d St., New York 17, N. Y. T/MS (FY-1962): $108,000,000. M/S and C/P: Predominantly liquid rocket fuels.

(43) *Bethlehem Steel Co.*, Bethlehem, Pa. T/MS (FY-1962): $99,900,000. M/S and C/P: Mainly structural steels and alloys for missile GSE, bulk rocketshells sheeting and other basic hardware.

(44) *Litton Industries Inc.*, 336 N. Foothill Rd., Beverly Hills, Calif. DIV: Aero Service, Airtron, Electron Tube, Ingalls Shipbuilding. T/P: 35,000. T/GR: CY-1961, $250,114,500; CY-1962, $280,-000,000; CY-1963 (est.), $305,000,000. T/MS (FY-1962): $89,-592,000 (NASA, $1,292,000; DOD, $88,300,000). M/S: Some aerospace electronics, but business is primarily shipping services (missiles from factory to various ports or sites), also heavy-duty GSE at launch sites.

(45) *White Motor Co.*, Springfield, Ohio. DIV: White Diesel Engines, Oliver Corp. T/MS (FY-1962): $87,400,000 (all DOD). M/S and C/P: Mainly Diesel electrical generating equipment, engines for mobile missile launchers, and heavy-duty motors for truck transport of rockets.

(46) *Kaiser Industries Corp.*, 300 Lakeside Dr., Oakland 12, Calif. DIV: Aircraft & Electronics, Aluminum & Chemicals, Electronics, Steel Fabricating. T/MS (FY-1962): $87,100,000 (all

DOD). M/S and C/P: Some electronics and chemical supplies, but mainly sheet steel, aluminum castings and light-alloy sheathing for rockets, satellites and spacecraft.

(47) *Burroughs Corp.*, 6071 Second Ave., Detroit 32, Mich. DIV: Burroughs Control. T/P: 40,000. T/GR: CY-1961, $401,210,-700; CY-1962, $424,680,000; CY-1963 (est.), $460,000,000 (missiles/space = 25%). T/MS (FY-1962): $86,800,000 (all DOD). M/S: Extensive line of computers, data-processors, and communications linkage. C/P: Computer systems for Atlas and Titan guidance, for satellite tracking radars, for cost analyses of space projects. Data-processing techniques for satellite telemetry information, simulated Saturn flight studies and space vehicle statistical R&D.

(48) *Continental Motors Corp.*, Market St., Muskegon 82, Mich. DIV: Aviation & Engineering, Gray Marine Motor, Wisconsin Motor. T/MS (FY-1962): $86,600,000 (all DOD). M/S: Jet engine components, rocket motor parts, missile electric-motor systems. C/P: Sub-contracts for air-breathing ramjet missiles (Hound Dog, Regulus I, Bomarc A). Variety of auxiliary motorized equipment for military rockets.

(49) *Kaman Aircraft Corp.*, Old Windsor Rd., Bloomfield, Conn. T/GR (CY-1962): $75,000,000; CY-1963 (est.), $82,000,000. T/MS (FY-1962): $83,800,000 (all DOD). M/S: Outside of helicopters and their motors for Army, various rocket-motor components. C/P: Study and R&D of aerodynamic recovery systems for big boosters and large spacecraft (based on helicopter techniques with unfolding blades, after burn-out and during descent).

(50) *Goodyear Tire & Rubber* Co., 1210 Massilon Rd., Akron 15, Ohio. DIV: Goodyear Aircraft Corp. (responsible for most aerospace work). T/MS (FY-1962): $83,200,000 (all DOD). M/S: Diversified range, including inflatable plastic radomes (protecting tracking radars), radio/radar antennas, filament-wound solid-fuel rocket casings, missile GSE, rubberoid linings. C/P: Prime for erectors of Atlas ICBM, radomes for Nike-Zeus radar complex, complete USN Subroc missile, GSE for Mace. Sub-contracting for inner rubberoid linings of Polaris and other missiles, "wire skids" landing gear of Dyna-Soar, MMRBM launcher GSE, Syncom tracking antennas. Also, NASA-funded studies of 23-man inflatable space station.

(52) *American Bosch-Arma Corp.*, Roosevelt Field, Garden City, N. Y. DIV: American Bosch-Arma, Tele-Dynamics. T/P: 9,800. T/GR: CY-1961, $133,636,000; CY-1962 (est.), $140,000,000; CY-1963 (est.), $150,000,000. T/MS (FY-1962): $81,400,000 (all DOD). M/S: Electronics, servo-hydraulics, guidance system com-

ponents. C/P: Sub-prime for USN's Terne missile, Titan-3 launcher, and Atlas ICBM missile site network. Sub-contracting in many other astro-electronics areas.

(55) *General Precision Equipment Corp.*, 50 Prospect Ave., Tarrytown, N. Y. DIV: GP Aerospace, GP Hydrospace, GP Systems, GP Research Center, GPL Division, Kearfott, Librascope, Link, Commercial Computer. T/P: 17,500 (3,900 technical). T/GR: CY-1961, $234,600,000; CY-1962 (est.), $222,862,000; CY-1963 (est.), $275,-000,000 (missiles/space = 77%). T/MS (FY-1962): $70,500,000 (all DOD). M/S: Doppler tracking techniques, stellar-inertial missile guidance, digital-analog computers, electromechanical servo-systems, global data-processors for ComSats, astro-electronics advanced research. C/P: Prime for MMRBM guidance. Sub-prime for guidance and fire-control of USN Subroc, Asroc, USAF Bomarc. Many tracking and data-handling computer modules throughout the missile and space rocket complex.

(58) *Aerospace Corp.*, 2400 E. El Segundo Blvd., El Segundo, Calif. This is the only *non-profit* organization to be listed: it plays a significant special role. As noted above, this government-controlled organization has taken over many of the former "prime-prime" duties TRW/STL once performed. Currently, only a few DOD missile projects are under STL's technical management, still fewer under MIT. Prime-prime services of NASA were turned over to BTL for the over-all manned flight program, and to GE/MSD for the reliability of the Apollo moonflight project. Aerospace Corporation, through NASA and the USAF/AFSC, is top prime over most other major space vehicle and military missile programs. T/MS (operating funds rather than "sales," FY-1962): $63,000,000. M/S: Mainly staffed with high-grade scientists, engineers and administrators, AC builds no hardware and dispenses services only—in basic research, in devising R&D plans for new projects, in acting as government/industry liaison, and in supplying on-site technical teams to oversee any given hardware contract. C/P: Prime-prime for complete Thor, Atlas and Titan strategic missile programs—manufacturing, testing and placement at operational sites. Also manages most tactical missile species—ground-to-air, air-to-ground, air-to-air, ground-to-ground, ship-to-shore, surface-to-underwater, etc. Such joint USAF/NASA projects as X-15, Gemini and Titan-3 are also under the control of Aerospace Corp. (Certain unsatisfactory results of this 2-year-old system, plus growing preference for industry "incentive" contracts—with bonuses for "good" work, penalties for "bad"—may curtail or modify the future role of AC.)

(71) *Laboratory For Electronics Inc.* ("LFE" in the trade), 1079 Commonwealth Ave., Boston 15, Mass. DIV: LFE Electronics, Eastern Industries, Automatic Signal, Tracerlab, Keleket X-ray, LFE International. T/P: 5,200 (25% technical, up 97% in 1961 and 70% in 1962). T/GR: FY-1962, $61,483,000; FY-1963 (est.), $70,000,-000. T/MS (FY-1962): $40,300,000 (all DOD). M/S: Strong in advanced physics research (mathematical physics, solid-state, plasma, astro-physics). Hardware includes gamut of electronic components for radar, tracking, guidance, computers, auto-controls, servo-magnetics, microwave devices. Also pumps, coolants, pressurization systems within missiles and space vehicles, electro-nuclear gadgetry, ray sensors, X-ray specialties, microcircuitry. C/P: Advanced electronic study contracts come steadily from the USAF, due to LFE's top-grade researchers in the omni-physics field. LFE is typical of the Class-A hardware suppliers, winning sub-contracts for innumerable components in telemetry, tracking, space guidance, missile control, miniaturized switch systems, nuclear space-power and other basic paraphernalia vital to earth/space rocketry.

(79) *Vitro Corporation of America*, 261 Madison Ave., New York 16, N. Y. DIV: Vitro Electronics, Vitro Labs, Vitro Engineering Lab, West Orange Lab. T/MS (FY-1962): $36,646,000 (NASA, $1,646,000; DOD, $35,000,000). M/S: In hardware, electronics specialties; in R&D services, engineered plans and detailed procedures for new astronautics projects. C/P: Prime for adding wideband telemetry to the Mercury-Gemini-Apollo tracking network; for instrumentation of Centaur launch-pad at Cape Canaveral; for gas-bag flight operations in Project Stratoscope-II (36-inch telescope lofted by balloon to 80,000 feet for sharp-focus astronomical photos of moon, planets and stars). Sub-prime in systems engineering for USN Talos, Tartar and Terrier missiles.

(84) *Ryan Aeronautical Co.*, 2701 Harbor Dr., San Diego 12, Calif. DIV: None. T/P: 5,125. T/GR: CY-1962, $72,508,800; CY-1963 (est.), $80,000,000. T/MS (FY-1962): $33,500,000 (all DOD). M/S: Leading designer of experimental VTOL (Vertical Take Off and Landing) jetplanes, including world's first VTOL, the X-13 Vertijet; also, Ryan's recent Verti-Plane is highly regarded by USAF. Other aerospace fields are inflatable vehicle studies (Flex-Wing re-entry glider for Saturn-booster recovery); orbit-unfolding solar mirror; advanced spacecraft metals (laminated steels, bonded metallo-ceramics); target-drones for military missiles; Doppler radar astrogation. C/P: Prime for Q-2 drone series as targets for military missiles. Sub-prime for velocimeter of lunar-landing Surveyors; for Saturn

booster accelerometers; for infra-red "docking" system of Gemini craft. Solar-cell panel beds for Ranger/Mariner/Surveyor series of interplanetary probes (including those aboard famed Mariner-2 during Venus flyby in December, 1962). Propulsion unit of Army's Corporal.

(94) *Texas Instruments Inc.*, P.O. Box 5474, 13500 N. Central Expressway, Dallas 22, Texas. DIV: None. T/P: 19,500. T/GR: CY-1961, $233,225,000; CY-1962, $240,692,500; CY-1963 (est.), $260,-000,000. T/MS (FY-1962): $29,959,000 (NASA, $959,000; DOD, $29,000,000). M/S: World's leading producer of semi-conductors (transistors, tunnel diodes, triodes). Pioneered missile-control electronics used widely in America's rockets, both military and civilian. C/P: Computer-programming systems for famed Thor-Delta booster; components of telemetry circuits for series of Mercury/Gemini/Apollo capsules; transistorized packs in Mariner and Surveyor vehicles; guidance gadgets in USN Shrike missile and USAF Minuteman.

(98) *Brown Engineering Co. Inc.*, P.O. Box 917, Huntsville, Ala. T/MS (FY-1962): $12,734,000 (NASA, $11,932,000; DOD, $502,000). M/S: Specializes in providing engineering plans for launch check-out, static-test systems, handling of rocket fuels, plus some GSE hardware in those areas. C/P: Center-of-gravity and mass-moment balance determinations for Saturn series of boosters; GSE for Saturn static-tests and Launch Complex #34 at Cape Canaveral; calibrations of inner instrumentation assemblies for NASA launch vehicles; DOD research projects into advanced GSE for Titan-2 and Titan-3 space launchers.

(99) *Electro-Optical Systems Inc.*, 125 N. Venido Ave., Pasadena, Calif. DIV: Fluid Physics, Solid State, Advanced Power, Advanced Electronics, Quantum Physics. T/MS (FY-1962): $8,215,000 (NASA, $852,000; DOD, $7,353,000). M/S: Mainly a high-grade research concern, with minimal hardware. World leader in R&D of "exotic" electric-propulsion for rockets (ion-drive, plasma-pulse, arc-jet). Also, optical-type space communications (via LASER); deep-space detection radars. C/P: Continually expanding research contracts from NASA/DOD to develop electric-ion spaceship propulsion, ship-to-ship lightbeam communications, advanced space-environment techniques (including zero-g simulations on earth), and "bio-chemical fuel cells" (human wastes converted thermally into power). Most DOD/USAF contracting is classified, dealing with space-defense "heat-ray" weaponry, super-velocity ion-drives (up to 25 mps) and "hostile satellite" inspection from manned military space stations. (Electro-Optical is representative of some 250 small but peerless frontier research facilities that are spearheading the next-decade space

technologies, of prime importance in winning the Space Race. Though modest in comparison with the lavish contracts of giant firms, Electro-Optical's dollar-volume of business in revolutionary research will pack a far more potent "Sunday punch" in the astro-technology of 1970.)

(100) *Arthur D. Little Inc.*, 35 Acorn Park, Cambridge, Mass. DIV: None. T/P: 1800 (90% research engineers and scientists). T/MS (FY-1962): $7,124,000 (NASA, $792,000; DOD, $6,332,000). M/S: Like Electro-Optical, Little's main product is "researchware," either pure or applied, conducted by a hand-picked team. C/P: Both DOD and NASA turn over to ADL their knottiest engineering problems in propellant-loading of giant rockets, missile-borne cryogenic systems, re-entry attack angles, welding of refractory metals that will not even start melting below 5000°, and other king-sized technological headaches. Wherever basic research has left a black void—in nucleonics, space medicine, zero-g phenomena, orbit decay of satellites—ADL works up enlightening breakthroughs with monotonous regularity. One of the oldest research organizations (founded 1886), ADL over 50 years ago actually made a silk purse out of a sow's ear, just to prove it could be done. Today, over 100 important basic patents are registered each year by this "invention mill." ADL is not the end of the Top Aerospace Hundred, but the beginning of the Thousand Aerospace Aces—small business firms producing a mountain of invisible mentalware without which our space effort would go two steps forward and three back. It would take another volume to do justice to these little giants of advanced science technology.

The firms omitted from the above listing of the Top Hundred are as follows in brief. Listed here are their relative position, total NASA/DOD aerospace contracts for FY-1962, and main contributions to America's astronautic adventure.

(51)	*International Harvester Co.*	$80,800,000	Heavy-duty GSE.
(53)	*The Magnavox Co.*	73,200,000	Electronics.
(54)	*Lear-Siegler Inc.*	73,100,000	Diversified instrumentation.
(56)	*Shell Oil Co.*	67,400,000	Rocket fuels.
(57)	*Socony Mobil Oil Co.*	65,800,000	Liquid propellants.
(59)	*Hayes International Corp.*	61,990,000	Diversified GSE.
(60)	*Morrison-Knudsen & Associates*	61,000,000	Diversified equipment.
(61)	*Bowen-McLaughlin-York Inc.*	59,000,000	Assorted bulk hardware.

(62)	*Eastman Kodak Co.*	55,400,000	Cameras and film for satellites and research rockets.
(63)	*Olin Mathieson Chemical Corp.*	53,300,000	Specialty fuels, plastics, bulk chemicals.
(64)	*Bath Iron Works Corp.*	51,200,000	Launch GSE and structural rocket hardware.
(66)	*Mason & HSM Co.*	44,300,000	Assorted hardware.
(67)	*Continental Oil Co.*	44,200,000	Rocket fuels.
(68)	*E. I. duPont de Nemours & Co.*	42,200,000	Bulk chemicals, plastics, synthetics, instrument casings.
(69)	*Hardeman-Fishback*	40,900,000	Diversified.
(70)	*Todd Shipyards Corp.*	40,400,000	Waterway rocket transportation.
(72)	*Kaiser-Raymond-Macco*	39,700,000	Varied GSE.
(73)	*Flying Tiger Line Inc.*	39,100,000	Air delivery of special instrumentation to launch sites.
(75)	*Harrington & Richardson Inc.*	38,500,000	Diversified.
(76)	*Richfield Oil Corp.*	38,400,000	Fuels and reagents.
(77)	*Motorola Inc.*	38,389,000	Electronics.
(78)	*Standard Oil Co. of Indiana (Amoco)*	36,800,000	Fuels.
(80)	*Hazeltine Corp.*	36,000,000	Radars and computers.
(81)	*Cities Service Oil Co.*	35,800,000	Rocket propellants.
(82)	*System Development Corp.*	35,100,000	Engineering studies and automated equipment for missiles.
(83)	*Fairchild Stratos Corp.*	35,000,000	Diversified missile/space components and systems.
(85)	*Motec Industries Inc.*	33,100,000	Assorted hardware.
(86)	*Blount Bros. Construction Co.*	32,148,000	Buildings, hangars, labs for space centers.
(87)	*ARO Inc.*	32,800,000	Diversified.

(88)	*Sanders Associates Inc.*	32,100,000	Electronics systems.
(89)	*Phillips Petroleum Co.*	32,000,000	Fuels.
(90)	*Universal American Corp.*	31,900,000	Diversified.
(91)	*Union Carbide Corp.*	31,586,000	Cryogenics (LOX), metals, plastics.
(92)	*Mitre Corp.* (non-profit)	30,200,000	Research, feasibility studies, R&D systems.
(93)	*Ogden Corp.*	30,200,000	Shipping services and GSE.
(95)	*Harvey Aluminum Inc.*	29,000,000	Structural rocket-shell metals.
(96)	*Gilfillan Corp.*	28,900,000	Electronics and instrumentation.
(97)	*Hallicrafters Co.*	28,700,000	Check-out systems and spaceflight simulation equipment.

Below the Top Hundred by dollar-volume only are a Technological Thousand whose service-volume far outweighs the contracts (from $25 million down to $500,000) they received in FY-1962 from the joint NASA/DOD space budget. To list the most outstanding: *Packard-Bell Electronics Corp.,* Calif. (space guidance); *Rust Engineering Co.,* Ala. (Saturn booster R&D); *Electronics & Missile Facilities Inc.,* N. Y. (telemetry); *Computdyne Corp.,* Pa. (cybernetics); *Radiation Inc.,* Fla. (ray sensors); *Ball Bros. Research Corp.,* Colo. (engineering breakthroughs); *Ampex Corp.,* Wash., D. C. (tape storage of space data); *Aero Geo Astro Corp.,* Va. (exotic aerospace research); *Geophysics Corp. of America,* Mass. (space weather data systems); *Kollsman Instrument Corp.,* N. Y. (miniaturized electronics); *High Vacuum Equipment Corp.,* Mass. (manned spaceflight simulators); *Telecomputing Corp.,* Calif. (space communications); *Hexcell Products Inc.,* Calif. (honeycomb re-entry heatshields); *Space Age Materials Corp.,* N. Y. (pyrogenics for spacecraft); *Electrospace Corp.,* N. Y. (tele-communications); *Hydro-Space Technology,* N. Y. (missile switches); *Rocket Research Corp.,* Minn. (advanced propulsion); *Stellardyne Labs Inc.,* Calif. (space environment); *Volt Tech Corp.,* N. Y. (rocket data documentation); *Zero Mfg. Co.,* Calif. (missile relays); *Astro-Science Corp.,* Calif. (magnetic tape in satellites); *Melpar Inc.,* Va. (synthetic ecology of astronaut food during long spaceflight); *Corning Glass Works*, N. Y. (space optics); *SKF Industries Inc.*, Pa. (micro-miniature gyro bearings); *Lionel Corp.*, N. J. (explosive bolts for rocket stage separations); *Cook Electric Co.*, Ill. (space telephonics); *Borg-Warner Corp.*, Ill. (astro-electronics).

A final salute to the other hundreds of space-specialist firms we must leave anonymous, but not forgotten. Their place, too, is assured in the unwritten astronautics annals yet to come.

And you, the science grads of this year, can bask in this pleasant thought—there are only 55,000 of you as against 200,000 aerospace firms shining up their help-wanted signs, a minimum of four jobs waiting for each of your trained brains. In the astronautics industry you are *Wanted Men*, dead(broke) or alive.

APPENDIX 5

*The Magnitude of the Federal
Space Budgets*

Missile/Space FY '64
Budget Request
(in millions of dollars)

Missile Procurement	4,108.0
Missile RDT&E	2,227.0
Missile Maintenance	1,500.0
Missile Base Construction	650.0
Polaris Submarines	702.3
NASA R&D	4,881.0
NASA Construction	785.0
Military Astronautics	1,667.6
AEC Astronautics	254.3
Weather Bureau Astronautics	26.2
NSF Astronautics	2.3
Associated Electronics	450.0
Other	1,370.0
TOTAL	**18,623.7**

How Missile/Space Spending

TABLE A   Government Spending for All Areas in the Field of Astronautics (actual and estimated) During a Four Year Period.

Source: *Missiles and Rockets*, Jan. 21, 1963

$11.38 $13.50 $16.22 $18.62

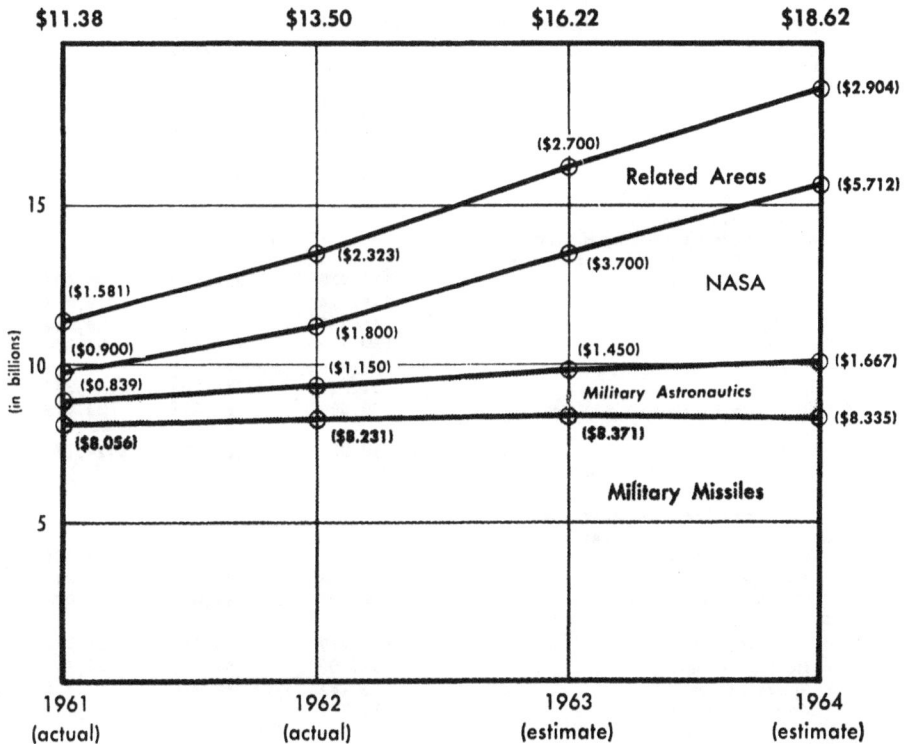

Has Climbed

TABLE B   The $5.7 Billion NASA Budget Requested for FY-1964
Source: *Aviation Week, Jan. 21, 1963*

Fiscal 1964 NASA Program Breakdown

(In thousands)

Program	FY1964	FY1963	FY1962
Manned Space Flight			
Research, development, operations			
Spacecraft development, operations.....	$1,647,441	766,696	193,192
Launch vehicles.....................	1,319,454	852,808	468,048
Supporting operations................	226,746	87,515	16,544
	$3,193,641	$1,707,019	$677,784
Facilities construction.................	564,538	535,091	245,418
Total.........................	$3,758,179	$2,242,110	$923,202
Unmanned space flight			
Research, development, operations			
Spacecraft development, operations			
Lunar, planetary....................	$331,270	226,392	164,631
Geophysics, astrophysics............	232,624	174,209	119,771
Bioscience.........................	41,339	25,112	4,082
Launch vehicles.....................	149,532	121,495	100,858
	$754,765	$547,208	$389,342
Facilities construction.................	25,509	55,817	39,829
Total.........................	$780,274	$603,025	$429,171
Space research technology			
Research, development, operations			
Launch vehicles, spacecraft			
Space vehicle systems..............	$111,407	89,758	58,869
Electronic systems.................	59,286	44,875	24,491
Human factors systems.............	24,387	14,075	4,996
Propulsion, space power			
Nuclear rockets....................	118,919	89,248	38,010
Nuclear electric....................	89,605	61,098	38,194
Space power.......................	25,184	18,316	12,256
Chemical propulsion................	35,077	25,119	17,164
	$463,863	$342,489	$193,980
Facilities Construction.................	74,979	95,436	37,815
Total.........................	$538,842	$437,925	$213,795
Support programs			
Research, development, operations			
Tracking, data acquisition............	$261,608	159,781	83,832
Training, research grants.............	56,438	31,666	13,547
	$318,046	$191,477	$97,379
Facilities construction.................	127,600	36,288	35,018
Total.........................	$445,646	$227,765	$132,397

Program	(In thousands) FY1964	FY1963	FY1962
Applications			
Research, development, operations			
Meteorological satellites..............	$73,085	64,289	39,696
Communications satellites............	55,771	49,049	35,591
Industrial applications................	6,099	3,760	877
Advanced applications..............	1,604	1,003	73
	$136,559	$118,101	$76,237
Facilities construction....................	4,103	....	....
Total..........................	$140,662	$118,101	$76,237
Aircraft technology			
Research			
Supersonic transport..................	$3,790	4,287	1,958
V/STOL aircraft.....................	1,600	1,037	182
X-15 aircraft......................	900	5,580	539
Other research, development..........	9,910	7,174	996
Institutional support..................	28,926	25,912	25,819
	$45,126	$43,990	$28,773
Facilities construction...................	3,271	6,697	110
Total..........................	$48,397	$50,687	$28,883

Fiscal 1964 Manned Space Flight Programs

Spacecraft Development, Opns.			
Apollo....................................	$1,207,400	$435,000	$75,618
Gemini....................................	306,300	232,838	54,959
Mercury....................................	....	21,299	31,060
Research....................................	42,900	18,600	5,164
Institutional Support......................	90,841	58,175	26,391
Total.............................	$1,647,441	$766,696	$193,192
Launch Vehicles			
Advanced Saturn	$733,000	$328,600	$57,375
Saturn.................................	131,600	228,615	215,320
Saturn B.................................	68,600	18,750	....
J-2 Engine................................	48,200	44,542	33,635
F-1 Engine................................	54,100	50,800	48,320
M-1 Engine................................	45,000	35,000	16,705
Research....................................	88,000	27,750	14,798
Institutional Support......................	150,954	118,752	81,895
Total.............................	$1,319,454	$852,809	$468,048
Supporting Programs			
Integration and Checkout.................	$157,378	....	....
Systems Engineering......................	47,528	....	....
Aerospace Medicine......................	21,840	....	....
Total.............................	$226,746	$87,515	$16,544
Construction..................................	564,538	535,091	245,418
Total, Manned Space Flight	$3,758,179	$2,242,110	$932,202

DEFENSE

TABLE C   The Status of Individual Space Programs Under the FY-1964 Budget.
Source: *Missiles and Rockets*, Jan. 21, 1963

ATLAS	Training missiles and equipment procurement continued, as well as retrofit and spares.
TITAN I	Training missiles and equipment procurement continued, as well as retrofit and spares.
TITAN II	Procurement to be completed.
TITAN III	Development continued as space booster workhorse.
MINUTEMAN	Production, installation and checkout continued, as well as retrofit and spares. Advanced version development continued. Additional buy of 150 for total of 950.
POLARIS	Procurement continued, and high priority assigned to improved version development. Support facilities to be built for both Atlantic and Pacific fleets. Missiles operational in 1966: 496.
SKYBOLT	Cancelled.
DYNA-SOAR (X-20)	Development continued.
ASROC	Procurement continued.
BMEWS	Support continued; reduction over FY '63 funding due to near-completion.
BULLPUP	Procurement increased for USN; procurement continued for USAF in both operational and trainer versions.
DAVY CROCKETT	Procurement continued.
DEFENDER	Support continued.
FALCON	Procurement continued.
FIREBEE	Procurement continued.
HAWK	Procurement continued.
HONEST JOHN	Procurement continued on improved versions.
HOUND DOG	More than 500 operational in 1966.
LANCE	Development continued.
LITTLE JOHN	Procurement continued on improved versions.
MAULER	Initial procurement and continued development.
MMRBM	Development planned if program definition phase satisfactory. DOD does not regard Polaris as alternative to either land- or sea-based versions.

NIKE-HERCULES	Procurement continued; sites to be improved and modified for fallout protection.
NIKE-ZEUS	Major R&D effort of $325 million, as well as initial development of a more advanced ballistic missile defense system. Extensive research on other techniques to be carried out by ARPA.
NIKE-X	Initial development to be undertaken at a high level of effort. Part of Zeus system.
PERSHING	Procurement continued.
REDEYE	Continuing R&D. No procurement due to development problems.
SAINT	Development continued on improved techniques.
SERGEANT	Procurement continued.
SHRIKE	Procurement increased for USN; procurement initiated for USAF.
SIDEWINDER	Procurement continued.
SPADATS	Support continued, facilities construction to continue.
SPARROW	Procurement continued for USN; procurement initiated for USAF.
SPRINT	Funded as complementary system to Nike-Zeus, Nike-X.
SUBROC	Procurement accelerated, development continued.
TALOS	Procurement continued in phase with related ship program.
TARTAR	Procurement continued in phase with related ship program.
TERRIER	Procurement continued in phase with related ship program.
TRANSIT	Development continued.
TYPHON	Development continued.
WALLEYE	Procurement to be initiated.

AEC

ROVER	Reactor development for NASA nuclear rocket program rises from $71 million to $98.9 million.
SNAP	Funding for reactor research for NASA program continues to rise, hitting $85.9 million.
PLUTO	R&D funding for DOD's nuclear-powered missile system cut back sharply from $22 million in FY '63 to only $9 million.

NASA

MERCURY — Last flight scheduled for April, 1963. No funds requested in FY 1964.

APOLLO — Funding to nearly triple—$1.2 billion tagged for Command, Service & Lunar Excursion modules.

GEMINI — Total requested authorization is $306.3 million, compared with $232.8 in the current year. First manned orbital flight will be attempted in 1964.

LUNAR LOGISTICS VEHICLE — No development funds requested.

MANNED ORBITING SPACE STATION — No hardware money included in budget request. Some funding of study contracts likely.

ADVANCED SATURN — Budget for 7.5-million-lb.-thrust booster up almost 200%—to $733 million.

SATURN — With first two-stage flight scheduled for mid-1963, development will drop to $131.6 million.

SATURN B — With development of improved one-J-2-engine stage just beginning, budget will jump from current $18 million to $68.6 million in FY '64.

LIQUID ROCKET ENGINES — Funding rising only slightly—$48.2 million for J-2, $54.1 million for F-1 and $45 million for M-1.

NOVA — No specific budget request, but part of $80-million launch vehicle R&D budget could be used for further study of giant booster.

TIROS — Funding slashed by two-thirds, to $7.2 million.

NIMBUS — Will get primary emphasis in weather satellite area, with funding of $43.8 million.

AEROS — No specific funding requested.

ECHO — Sharp drop-off—to only $200,000—as passive ComSat program is ended.

RELAY — Funding slashed to $1.9 million as DOD takes over responsibility for low-altitude system.

ADVANCED RELAY — No funds requested. Project apparently killed.

REBOUND — No funds requested. May continue as R&D project.

SYNCOM — Budget drops to $4 million as emphasis moves to advanced system.

ADVANCED SYNCOM — NASA now has prime responsibility for development of high-altitude ComSat. Budget jumps to $40 million.

ADVANCED APPLICATIONS SATELLITES — Only $1 million requested, perhaps for study of navigational and data-gathering satellite systems.

INDUSTRIAL APPLICATIONS — Budget calls for only $3.5-million program to make NASA technology available to industry.

SCIENTIFIC OBSERVATORIES — Funding to $132 million requested for OGO, OSO and OAO.

SCIENTIFIC SATELLITES Budget rises slightly—to $26.8 million—for Earth satellites.

INTERNATIONAL PROGRAMS Cut back from $12 million to $7 million.

RANGER Lunar hard-landing spacecraft budget increases slightly, to $90 million.

SURVEYOR LANDER Funding set at $69.3 million in effort to meet 1965 launch date.

SURVEYOR ORBITER Gets first major funding support, with 14-fold increase to $28.2 million.

MARINER Budget almost doubles, to $100.1 million for 1964-65 interplanetary missions.

VOYAGER No R&D funds requested. Some study contracts may be awarded.

PIONEER New solar probe program kicked off with $15-million request.

BIOS Flight program to orbit squirrel monkeys and other life forms rises sharply—from $10.9 million in current fiscal year to $24.2 million.

SCOUT R&D completed. Procurement funds included in other programs.

DELTA R&D completed. Procurement funds included in other programs.

CENTAUR High-priority upper-stage program funding rises slightly, to $110.7 million.

RE-ENTRY PROJECTS Funding remains level—at $14 million—for Project Fire and other programs.

MICROMETEOROID
SATELLITES New 4000-lb. satellite program earmarked for $5.2 million.

SNAP-8 Funding rises from $15.9 million in FY '63 to $24 million.

SERT R&D program increases by 300%, to $15.3 million.

NERVA Nuclear rocket engine program budget continues to rise, to $60 million.

RIFT Funding for nuclear rocket stage stays at low ebb with $12-million request.

WEATHER BUREAU

METEOROLOGICAL
SATELLITES Procurement and launch of Nimbus satellite budgeted at $23 million. Another $20.6 million in unobligated FY '63 funds also available.

APPENDIX 6

The Future of America's Space Program:
the Plans and the Hardware

TABLE A   Long-Range U.S. Plans for Space Exploration

Source: *Aviation Week*, Nov. 12, 1962

MISSIONS	REGION		
	EARTH ORBIT	**LUNAR**	**PLANETARY**

UNMANNED	1958 – 1
	UNMANNED SATELLITES
	
	SCIENTIFIC SATELLITES
	✻SMALL SPECIAL PURPOSES
	✻ORBITING OBSERVATORIES
	
	APPLICATION SATELLITES
	✻COMMUNICATION
	✻METEOROLOGY
	✦NAVIGATION
	✻ENGINEERING RESEARCH
		1962 – 68 2
		LUNAR PROBES
		✻RANGER
		✷SURVEYOR
		
		INTERMEDIATE
		
		SPACE PROBES
			1962 3
			DEEP SPACE PROBES
			✻MARINER
			VOYAGER
			SEARCH FOR EXTRATERRESTIAL
			   LIFE
			
			OUT OF ECLIPTIC
			GRAVITATIONAL EXPERIMENT
			OUTER PLANETS AND THIER
			   SATELLITES
			LEAVE SOLAR SYSTEM

MANNED DEVELOPMENTAL	1962 – 68 4 **MANNED SATELLITES** BALLISTIC REENTRY ✷MERCURY ✦GEMINI: MANEUVERING REENTRY INTERIM ORBITAL LABS.	BEFORE 1970 5 **MANNED LANDING** ✦APOLLO LUNAR LOGISTIC SYSTEM ✷ ✦ UNMANNED	AFTER 1975 6 **MANNED EXPEDITIONS** MARS LANDING VENUS RECONNAISSANCE SEARCH FOR LIFE ON PLANETS NO AUTHORIZED PROGRAMS YET
MANNED OPERATIONAL	AFTER 1968 7 **ORBITAL OPERATIONS** MANNED ORBITING LABS. OPERATIONAL FERRY VEHICLE RECOVERABLE BOOSTERS ENGINEERING EXPERIMENT AND DEVELOPMENT NO AUTHORIZED PROGRAMS YET	AFTER 1970 8 **LUNAR STATION** SCIENTIFIC OBSERVATIONS LUNAR EXPLORATIONS NO AUTHORIZED PROGRAMS YET	AFTER 1980 9 **PLANETARY OPERATION** MARS STATION ADVANCED MANNED EXPEDITIONS JUPITER SATELLITES MERCURY OTHERS NO AUTHORIZED PROGRAMS YET

Note: Stars indicate programs approved as of time of writing. Projects in boxes 6 and 9 are under preliminary study, and are not approved or funded.

TABLE B   The Apollo project for a manned lunar landing before 1970
is currently America's biggest space exploration venture.
Here is how it has been set up.

Source: *Missiles and Rockets*, May 21, 1962

The Apollo Team

Center	Responsibility	Chief Procurement Officer
NASA Headquarters	Overall Management Responsibility	Ernest W. Brackett
Manned Spacecraft Center	Spacecraft	David W. Lang
Marshall Space Flight Center	Launch Vehicles	W. A. Davis
Launch Operations Center	Mission launches & Support	Gerard Michaud

Launch Vehicles

Vehicle	Contractor	Amount
Saturn C-1		
S-1 Stage	Chrysler Corp.	
S-IV Stage	Douglas Aircraft Corp.	
Advanced Saturn		
S-1B Stage	Boeing Co.	$500 million
S-II Stage	North American Aviation, Inc.	$300 million
S-IVB	Douglas Aircraft Corp.	$200 million
Engines		
H-1 (Saturn C-1)	Rocketdyne Div., NAA	
F-1 (Adv. Saturn, Nova)	Rocketdyne Div., NAA	$200 million
J-2 (Adv. Saturn, Nova)	Rocketdyne Div., NAA	$150 million
M-1 Nova	Aerojet-General Corp.	$130 million
RL10A3 (Saturn C-1) (Centaur)	Pratt & Whitney Aircraft	

The Line-Up of Apollo Contractors

Company Name and Address	System	Announce-ment Date	Amount
Apollo Spacecraft			
North American Aviation, Inc. Space & Information Systems Div. 12214 Lakewood Blvd. Downey, Calif.	Command & Service Modules	1961	$3 to 5 billion
Subcontractors			
AiResearch Manufacturing Co. 6201 West Imperial Highway Los Angeles, Calif.	Environ-mental control	12/21/61	$10 M
Avco Corporation Wilmington, Mass.	Heat shield	3/23/62	$8 M+
Collins Radio Company 855 - 35th Street, N.E. Cedar Rapids, Iowa	Tele-communi-cations	12/21/61	$40 M+
Lockheed Propulsion Co. P.O. Box 111 Redlands, Calif.	Escape motor	2/13/62	$5 M
Marquardt Aircraft 16555 Saticoy Van Nuys, Calif.	Reaction control rocket engines	3/2/62	Undeter-mined
Aero-Minneapolis Div. Minneapolis-Honeywell 2600 Ridgeway Rd. Minneapolis 8, Minn.	Stabiliza-tion and flight con-trol system	12/21/61	$30 M+
Pratt & Whitney Div. of United Aircraft East Hartford, Conn.	Fuel cell	3/9/62	
Ventura Div. Northrop Corp. 8000 Woodley Van Nuys, Calif.	Parachute landing system	12/21/61	$1 M+

Company Name and Address	System	Announce-ment Date	Amount
Thiokol Chemical Corp. Hunter-Bristol Div. P.O. Box 27 Bristol, Pa.	Tower jettison motor	4/6/62	$1 M+
Aerojet-General Space Propulsion Div. Azusa, Calif.	Service module propulsion engine	5/3/62	$12 M

Guidance & Navigation System

MIT Instrumentation Lab. Cambridge 39, Mass.	Manage-ment of G&N	8/ /61	$20 million
AC Spark Plug Div. General Motors Corp. Milwaukee 1, Wis.	Inertial Platform & Assoc. GSE	5/8/62	$16 million
Raytheon Company Bedford, Mass.	Guidance Computer (on-board)	5/8/62	$2 million
Kollsman Instrument Co. Elmhurst, N.Y.	Optical Subsystems (G&N)	5/8/62	$2 million
Sperry Gyroscope Div. Sperry Rand Corp. Great Neck, L.I., N.Y.	Acceler-ometers (G&N System)	2/23/62	Est. $.8 million

TABLE C   United States Space Booster Rockets

Source: Aerojet-General *Spacelines and Rocket Review,* 1962

United States Space Booster Rockets

ROCKET	STAGES	THRUST 1,000'S OF POUNDS	MAX. DIA. (FEET)	HGT. LESS SPACE-CRAFT (FEET)	PAYLOAD (LBS.) 345 MILE ORBIT	PAYLOAD (LBS.) EARTH ESCAPE
Scout	**Four stages	98, 48, 13.6, 2.8	3.3	65	150	—
Delta	Thor **Able Altair	150 7.5 2.8	8.8	77	500	60
Thor-Ablestar	Thor **Ablestar	150 7.8	8.8	83	800	—
Thor-Agena B	Thor Agena B	150 16	8.8	80	1,600	—
Atlas-Agena B	Atlas D Atlas D Sustainer Agena B	367 80 16	10	98	5,000	750
Titan II	**LR-87 **LR-91	430 100	10	90	—	—
Centaur	Atlas D Atlas D Sustainer *Centaur (2 A-3 engines)	367 80 30	10	105	8,500	2,300
Saturn	S-I (8 H-1 engines) *S-IV (6 A-3 engines)	1,500	22	125	20,000	—
Advanced Saturn	S-IB (5 F-1 engines) S-II (5 J-2 engines) *S-IVB (One J-2)	7,500 1,000 200	33	275	200,000	85,000
Nova	N-I (8 F-1 engines) **/*N-II (4 M-I engines) *N-III (One J-2)	12,000 4,800 200	50	280	350,000	150,000

*USES HIGH-ENERGY LIQUID HYDROGEN-LIQUID OXYGEN.
**AEROJET MANUFACTURES SCOUT FIRST STAGE, ABLE, ABLESTAR, TITAN II ENGINES, NOVA N-II STAGE ENGINES.
PROPELLANTS: ALL LIQUID EXCEPT SCOUT AND DELTA ALTAIR STAGE—SOLIDS.

TABLE D The Saturn and Nova rockets that will be used in the Apollo project.

EARTH ORBIT
SATURN C-1

20,000 LBS.
300 MILE ORBIT

SPACECRAFT

2ND STAGE
6 A-3 ENG.

1ST STAGE
8 H-1 ENG.

CIRCUMLUNAR
ADVANCED
SATURN

100,000 LBS.
300 MILE ORBIT

SPACECRAFT

3RD STAGE
1 J-2 ENG.

2ND STAGE
5 J-2 ENG.

1ST STAGE
5 F-1 ENG.

LUNAR LANDING
(RENDEZVOUS)
ADVANCED SATURN

200,000 LBS.
300 MILE ORBIT

SPACECRAFT

3RD STAGE
1 J-2 ENG.

2ND STAGE
5 J-2 ENG.

1ST STAGE
5 F-1 ENG.

LUNAR LANDING
(DIRECT)
NOVA

400,000 LBS.
300 MILE ORBIT

SPACECRAFT

3RD STAGE
1 J-2 ENG.

2ND STAGE
4 M-1 ENG.

1ST STAGE
8 F-1 ENG.

APPENDIX 7
Additional Sources for Student-Aid and Career Information

PART I. FREE OR LOW COST LITERATURE

GPO refers to the Government Printing Office, Washington 25, D. C. When ordering from the GPO, give full title, date, author or agency and catalog number, if any, for each item.

Accredited Technical Institute Programs in the U. S. New York: Engineers Council for Professional Development, 29 West 39th St., New York 18. 25¢. Published annually.

Career Guidance Kit for Engineering Technicians. Washington: National Council of Technical Schools, 1507 M St., N.W., Washington 5. 50¢. A group of 16 pamphlets about various technicians' jobs.

Careers in Astronautics. New York: American Institute of Aeronautics and Astronautics, 500 Fifth Ave., New York 36, N. Y. Free booklet.

Education Directory, Part III: Higher Education. Revised annually. Washington: Dept. of HEW, Office of Education. Order from GPO. 75¢. Register of U. S. colleges and universities.

Encouraging Future Scientists: Key to Careers. Washington: National Science Teachers Association, 1201 16th St., N.W., Washington 6. Free booklet.

Engineering, A Career of Opportunity. Washington: National Society of Professional Engineers, 2029 K St., Washington 6. Free 16-page pamphlet.

The Engineering Technician. Urbana, Ill.: American Society for Engineering Education, University of Illinois. 25¢. A booklet.

Financial Aids for Undergraduate Students, Sources of Information. Washington: GPO, 1962. Catalog No.: FS-5.255:55029. 9 pages. 10¢.

Financial Assistance for College Students: Undergraduate. Washington: GPO, 1962. Catalog No.: FS-5.255:55027. 360 pages. $1.25.

Guide to Scholarships, Fellowships and Loans. By John Bradley. New York: New American Library, a Signet Book (501 Madison Ave., New York, N. Y.) 75¢.

Guidebook for Parents. Columbus, Ohio: *Battelle Memorial Institute.* Free. Booklet to help parents discover and encourage future scientists and engineers.

Higher Education: Basic Student Charges (1961-62). Booklet OE-52005-62. Washington: Dept. of HEW, Office of Education. Order from GPO. 40¢.

How to Choose Your Technical Institute. By Walter M. Hartung and G. W. Brush, Jr. Cambridge, Mass.: Bellman Publishing Co., Cambridge 38. $1.00.

Information on Science Scholarships and Student Loans. Washington: National Science Foundation, June 1960. 9 pages. 15¢.

Job Horizons in Missile and Spacecraft Manufacturing; and *Mathematics and Your Career.* Washington: U. S. Dept. of Labor, Bureau of Labor Statistics, Occupational Outlook Service, Washington 25. Free booklets.

National Defense Student Loan Program: Student Borrowers, Their Needs and Resources. Booklet OE-55011. Washington: Dept. of HEW, Office of Education. Order from GPO. 55¢.

Navy Airmen: Space Age Specialties. Washington: U. S. Navy Dept., Bureau of Personnel (Room 1833, Navy Annex), Washington 25. Free leaflet.

Need A Lift? Indianapolis: *The American Legion,* Dept. S, P.O. Box 1055, Indianapolis, 6. 25¢. Annual revised 100-page booklet of career and scholarship information.

Pocket Guide to Air Force Opportunities. Write to: Commander, USAF Recruiting Service, Wright-Patterson AFB, Ohio. Free USAF booklet.

Projects: Space. By Judith Viorst (under NASA and Science Service sponsorship). New York: Washington Square Press, Inc., 1962. 45¢.

Summer Jobs for Students. Washington: U. S. Dept. of Labor, Bureau of Employment Security, Washington 25. 4 pages. Free.

Your Career in Industry as a Scientist or Engineer. Burlington, Iowa: National Research Bureau, Inc., 424 N. 3d St. 32 pages. 20¢.

Your Opportunities in Industry as a Technician. New York: National Association of Manufacturers, 2 East 48th St., New York 17. Free booklet.

PART II. CAREER BOOKLETS IN SERIES

Career Briefs. Large cards that list on both sides over 150 different occupations. 25¢ each. For full list, write to: Careers, Largo, Florida.

Modern Vocational Trends Career Monographs (i.e., *Careers in Astronautics and Space Exploration*, etc.). Paperbound books. $1.25 to $1.50 each. Obtain list from: World Trade Academy Press, Inc., 50 East 42d St., New York 17, N. Y.

Should You Be a Physicist? etc. Free booklet. For full list, write to: New York Life Insurance Co., Career Information Service, Box 51, Madison Square Station, New York 10, N. Y.

Visual Career Guides (i.e., *Your Career Opportunities in Engineering*, etc.). About 65 pages each. 75¢ per copy. For full list, write to: Rowman & Littlefield, Inc., 84 Fifth Ave., New York 11, N. Y.

Vocational and Professional Monographs. Booklets about scholarships, careers, science professions, etc. $1.00 each. Request list from: Bellman Publishing Co., Cambridge 38, Mass.

PART III. COMPREHENSIVE BOOKS IN DEPTH

Adams, Carsbie C., Wernher von Braun and Frederick I. Ordway III. *Careers in Astronautics and Rocketry.* New York: McGraw-Hill Book Co., Inc., 1962. 252 pages. $6.95.

Coombs, Charles I. *Rocketmen: What They Do.* New York: Franklin Watts, Inc., 1962. 184 pages. $3.95.

Ely, Col. Lawrence D. (USAF, ret.). *Your Future in Aerospace Technology.* New York: Richards Rosen Press, Inc., 1962. 153 pages. $2.95.

Eskow, Seymour. *Barron's Guide to the Two-Year Colleges.* Great Neck, N. Y.: Barron's Educational Series, Inc. Paperbound. $2.98. A complete résumé—in over 400 pages—of all junior, community, vocational and technical colleges or institutes. Regularly updated; price subject to change.

Lovejoy, Clarence E. *Lovejoy's College Guide.* New York: Simon and Schuster, Inc. A complete (over 500 pages) paperbound reference to all U.S. colleges and universities. Regularly revised; price subject to change. $2.50.

Markets of the Sixties. By the Editors of *Fortune.* New York: Harper & Row, Publishers, 1960. 266 pages. $5.00.

Merrill, Arthur C. *Investing in the Scientific Revolution.* Garden City, N. Y.: Doubleday & Co., Inc., 1962. 293 pages. $5.95. This book, and the one following, are recommended even though they are not strictly in the "career or student-aid" category, because both books give an enormous new insight into the aerospace/astronautics industry of today and tomorrow, and are thus of value to the technical student as studies-in-depth of the field in which his future lies.

———, *Careers and Opportunities in Science: A Survey of All Fields.* Rev. ed. New York: E. P. Dutton & Co. $3.95.

Nourse, Alan E. *So You Want to Be a Scientist?* New York: Harper & Row, Publishers, 1962. $3.00.

———, and J. C. Webbert. *So You Want to Be an Engineer?* New York: Harper & Row, Publishers, 1962. $3.50.

Pollack, Philip. *Careers and Opportunities in Engineering.* New York: E. P. Dutton & Co., 1962. 140 pages. $3.75.

The Space Industry, America's Newest Giant. By the Editors of *Fortune.* Englewood Cliffs, N. J.: Prentice-Hall, Inc., 1962. 192 pages. $3.95

The following series of books is recommended as "inspiration" reading for the aspiring young space scientist or astro-engineer. Each volume gives excellent biographical sketches of ten "space VIP's" and the feats they performed, which will be part of space history for all time to come.

Thomas, Shirley. *Men of Space*. Philadelphia: Chilton Co.

Volume I: Krafft Ehricke, Wernher von Braun, Robert Goddard, James Van Allen, others. 1960. $3.95.

Volume II: William Pickering, Scott Crossfield, Robert Truax, Hugh Dryden, others. 1961. $3.95.

Volume III: Alan Shepard, Yuri Gagarin, Donald Flickinger, James Doolittle, others. 1961. $3.95.

Volume IV: John Pierce, Hubertus Strughold, Robert Gilruth, others. 1962. $3.95

Volume V: John Glenn, Homer Newell, Robert Seamans, L. Eugene Root, others. 1962. $4.95.

PART IV. COMPREHENSIVE LISTINGS

(Bibliographic compilations of titles relating to every phase of scholarships, colleges, curricula, careers, jobs, aerospace industries and the space program.)

Aeronautics and Space Bibliography, for College Grades and Adults. NASA EP-3. 30¢. Compiled by NAEC. Order from GPO.

Aeronautics and Space Bibliography, for Elementary Grades. NASA EP-1. 25¢. Compiled for NASA by NAEC. Order from GPO.

Aeronautics and Space Bibliography, for Secondary School Grades. NASA EP-2. 30¢. Compiled by NAEC. Order from GPO.

Careers in Science. Washington: American Association for the Advancement of Science, 1515 Massachusetts Ave., N.W., Washington 6. 21 pages. 15¢. Lists publications of aid to high-school students in planning a college education leading to a scientific career.

Careers in Science, Mathematics and Engineering: A Selected Bibliography. OE-26007. Washington: U. S. Dept. of HEW, Office of Education, Washington 25. Free.

Education. 50th ed., July, 1962. Price List No. 31. Order (free) from GPO. Lists all GPO pamphlets available in this field.

Selected Aerospace Career and Scholarship Information. Circular OE-26005-1A-569B. Revised 1962. 16 pages. Free. Request from: Specialist for Aerospace Education, U. S. Dept. of HEW, Office of Education, Washington 25,.D. C.

Student Financial Aid in Higher Education. GPO Catalog No. FS-5.253:53006. 1961. 87 pages. 30¢. Annotated booklet.

PART V. ORGANIZATIONS OFFERING CAREER AND SPACE AGE INFORMATION

(Special questions and problems of the student or teacher can often be answered through these non-profit agencies.)

Aerospace Industries Association of America Inc.
1725 DeSales St.
Washington, D. C.

National Aeronautics and Space Administration
Office of Educational Programs and Services
Washington 25, D. C.

National Aerospace Education Council
1025 Connecticut Ave., N.W.
Washington 6, D. C.

National Science Teachers Association
1201 16th St., N.W.
Washington 6, D. C.

Thompson-Ramo-Wooldridge, Inc.
Public Information Office
8433 Fallbrook Ave., Canoga Park
Los Angeles, Calif.

(This well-known aerospace firm provides a special information service to the science or engineering student for his theme or school project. Questions on any current program in space, missiles, electronics or astronautics will be answered.)

Almost all other aerospace firms mentioned in this book also supply Space Age booklets, in-house scholarship information, lists of current technical help needed, and other student/job-seeker aids. All are *free*. Here are some typical samples, choosing only a few:

Can I Be a Technician?—General Motors Corp.

Computing Engineering—Douglas Aircraft Co.

Partners in Space—Aerojet-General.

The Space Age—Bell Telephone Co. (AT&T).

(For a list of their available free literature, write to the Public Relations Office of any of the firms previously reviewed in this book, using the *corporate headquarters* address given.)

The armed services also supply voluminous literature and data about their Space Age activities, technical training schools and many job opportunities for enlisted men. If interested, write to:

Air Force Association
Aerospace Education Council
Mills Building
Washington 6, D. C.

Department of the Air Force
Office of Information
Washington, D. C.

Department of the Army
Office of Information
Washington, D. C.

Department of Defense
Office of Defense Research and Engineering
Washington, D. C.

Department of the Navy
Office of Information
Washington, D. C.

Federal Aviation Agency
Office of Information
Washington, D. C.

Also, certain government agencies can be consulted for the many invaluable student-aid and career-guidance services they provide. Contact the following (always stating precisely what information you wish):

National Education Association
Career and Student Aid Information
Washington 25, D. C.

National Science Foundation
Educational Programs
Washington 25, D. C.

Occupational Outlook Service
(or Youth Employment Service)
(or Education Information Service)
U. S. Department of Labor
Washington 25, D. C.

U. S. Office of Education
Department of Health, Education and Welfare
Washington 25, D. C.

Another group of non-profit organizations, which supplies a variety of worthwhile services to the technical student, includes societies of technical professionals, such as:

American Association for the Advancement of Science
1515 Massachusetts Ave., N.W.
Washington 5, D. C.

American Institute of Aeronautics and Astronautics
500 Fifth Ave.
New York 36, N. Y.
(Formerly the American Rocket Society and the Institute of Aerospace Sciences, but now merged)

Engineers Joint Council
345 East 47th St.
New York 17, N. Y.
(At the same address: JETS [Junior Engineering Technical Society], Engineering Manpower Commission)

National Society of Professional Engineers
2029 K St., N.W.
Washington, D. C.